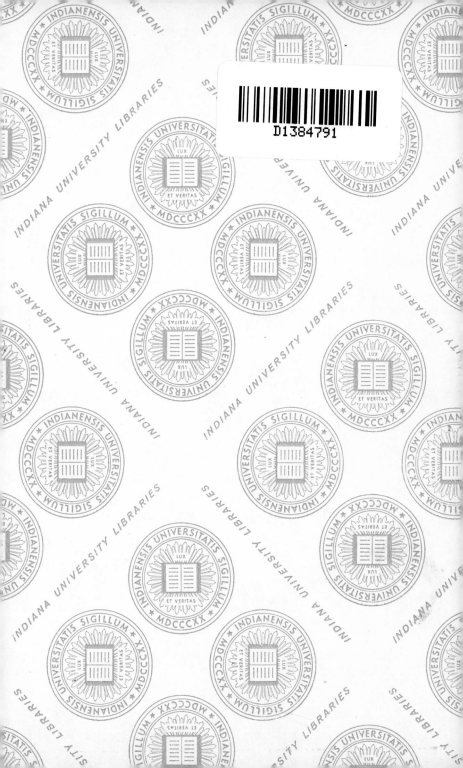

The Names and Structures of Organic Compounds

The Names and Structures of Organic Compounds

by Otto Theodor Benfey

Earlham College

QD 257
.B4

QD 257
.B4

John Wiley & Sons, Inc.   New York • London • Sydney

THE HECKMAN BINDERY, INC.   N. MANCHESTER, IND.

Copyright © 1966 by John Wiley & Sons, Inc.

All rights reserved. This book or any part thereof must not be reproduced in any form without the written permission of the publisher.

10   9   8

INDIANA
UNIVERSITY
LIBRARY

NORTHWEST

ISBN 0 471 06575 7
Library of Congress Catalog Card Number: 66-16550
Printed in the United States of America

To Stephen, Philip, Christopher, and Karen

May their educational experience in the future be more

enjoyable as well as more efficient as a result of the

many educational experiments currently being tested

Preface

This is a programmed book in a limited field. It deals with the common names of certain organic compounds and with the official "systematic" names assigned to them by the International Union of Pure and Applied Chemistry. It deals also with "structural formulas" of compounds, "structural" being used in a technical sense, meaning order of connection of the atoms in a molecule. The spatial arrangement of the atoms — the realm of stereochemistry — is not covered in this book.

Few persons other than those studying organic chemistry need to know names and structures of organic compounds. This program is, therefore, designed to be used in conjunction with a course in organic chemistry. It should permit the complete omission of nomenclature discussions from lectures in the introductory organic course. On the other hand, there may be individuals who are fascinated by the construction of a systematic system of nomenclature, more as a coding and decoding problem than as a scientific tool. The book is accordingly written in such a way as to permit its being mastered without a parallel course in organic chemistry. It should also be useful to librarians, editors and writers of technical material who have not had a scientific training.

The later chapters may seem to stray rather far from the title of the book, yet the topics are all closely related to the understanding of structures and formulas. Formal charge, electronic formulas, resonance structures, are topics mastered only by practice. The writer has always enjoyed lecturing on these subjects, yet he has painfully discovered that his students absorbed very little, that the seemingly successful lectures in fact failed. Furthermore, by the time he made the discovery, through problem assignments or examinations, he was well into another lecture subject and it did not seem opportune to cover the previous topics a second time. Programmed instruction informs the student, at the time he is studying, whether or not he is catching on. It is, therefore, an ideal mode of instruction for closely reasoned subject matter developed from the simple to ever more complex problems and applications. If the simpler steps are not understood, the student will know it before he dives into the advanced topics.

The final chapter is icing on the cake. How does one determine the fact that there are 75 possible structural isomers of formula $C_{10}H_{22}$? How does one write the 75 without danger of repeating an earlier structure in a slightly different arrangement? The chapter develops a method first published in 1931 by H. R. Henze and C. M. Blair. In the process of working through this chapter the student will review from a new viewpoint much of the material in earlier sections.

The book follows the programmed instruction technique known as linear programming. The student is presented information in small segments, given questions on each point, required to respond, and enabled to check immediately the correctness of his response. In this program, the student always moves forward in a straight line; no loops or branches have been incorporated in it. It can be used by an instructor in at least three ways:

a) in the form of pre-class assignments
b) in class
c) as post-class review.

I have used the material mainly in the first manner, giving up completely any lecturing on the principles of nomenclature, and proceeding on the assumption that the names used in class to discuss chemical reactions are understood by the student. This procedure has been effective and has liberated valuable lecture time.

If the program is used in class, the self-pacing factor of programmed instruction takes effect. The instructor has the advantage of being able to take care of difficulties as they arise for individual students, but he is also faced with the problem of what to do with students who finish early or those who do not finish within the allotted time.

No doubt students have also used the programmed chapters for review purposes, although the tabulations now appearing at the end of this book may serve that purpose more effectively. The reader interested in the experience of the Earlham Chemistry Department with a variety of programmed instruction materials, including those in this book, may wish to consult the report by G. R. Bakker, O. T. Benfey, and W. J. Stratton in the *Journal of Chemical Education* 40, 20 (1963).

Although the program was developed through a long process of testing with students, revising, and retesting, some rough spots no doubt have remained. Since programming is still very much in the experimental stage, criticisms and suggestions for improvement would be very much appreciated. So, too, would the results of controlled tests, comparing the program with other modes of instruction, or, better still, comparing various uses of the program within a curriculum.

I owe thanks to Dr. Kurt L. Loening, Director of Nomenclature, Chemical Abstracts Service, Columbus, Ohio, for reviewing those sections of the book dealing with systematic nomenclature and giving me the benefit of his expert knowledge. Thanks are due also to Lucille Rice and Kate Korfer for the not inconsiderable task of typing several versions of this program and to John Barlow, John Gilpin and M. Daniel Smith for introducing me to programmed instruction techniques. Many friends and colleagues have commented on early versions of the program and their help is gratefully acknowledged. Some of the early chapters were first developed under a grant to Earlham College from the United States Office of Education, Department of Health, Education and Welfare (No. 7-12-026.00) in support of college-wide explorations of programmed instruction.

O. Theodor Benfey

Richmond, Indiana
January, 1966

To the Student

In this book the answer will always be found following the black horizontal line that indicates the end of the question frame. To work through the program, cover the page you are reading with the answer shield provided with this book and move the shield down until the first horizontal line becomes visible. Now read the question above the line, answer it by writing in the space provided, and check your answer against the printed one appearing below the horizontal line. Move on to the next question and proceed in the same way.

Contents

The Names and Structures of Organic Compounds

The Common Names of Simple Organic Compounds

PART A—SATURATED ALIPHATIC HYDROCARBONS

Introduction: A *hydrocarbon* is a molecule made up of carbon atoms, hydrogen atoms and no other atoms.

1-1   In a hydrocarbon with n carbon atoms, the *maximum* number of hydrogen atoms is found to be given by the formula

maximum number of H atoms = $2n + 2$

If a hydrocarbon contains only one carbon atom, then

$(2 \times 1) + 2$ or _____ is the _____
(number)
number of hydrogen atoms.

---

4   maximum

1-2   A hydrocarbon with the maximum number of hydrogen atoms is called an alkane.
Methane, or marsh gas, is an alkane. Its formula is CH_4 (one carbon, four hydrogens).
Ethane is an alkane; it has two carbon atoms, so its formula is C_2H___ .

---

6 (i.e., C_2H_6)

1-3   Propane is an alkane with three carbons; its formula is _____.

C_3H_8

1-4   A hydrocarbon with the formula C_4H_{10} belongs to
the class of _____ because it contains the
_____ number of hydrogen atoms, given by
the formula $C_n H$_____.

---

alkanes   maximum $2n + 2$ (i.e., $C_n H_{2n+2}$)

1-5   When we reach alkanes with four or more carbon
atoms, we find that more than one compound has a
given formula. Thus *two* different chemicals with
four carbons and belonging to the alkanes exist. Both
have the formula _____.

C_4H_{10}

1-6   In the one case, the four carbons are in a chain,
$C-C-C-C$, whereas, in the other, only three are
in a chain, the fourth being attached to the center of
the chain, $C-C-C$. Carbon has a valence of 4 and
                                  |
                                  C

hydrogen is monovalent; a carbon atom forms four
bonds with other atoms and hydrogen can only link
to one other atom by one bond. When we attach hydrogen atoms to all the unused carbon valences, the four-
carbon chain formula can be represented as

$$H-\overset{\displaystyle \overset{H}{|}}{\underset{\displaystyle \underset{H}{|}}{C}}-\overset{\displaystyle \overset{H}{|}}{\underset{\displaystyle \underset{H}{|}}{C}}-\overset{\displaystyle \overset{H}{|}}{\underset{\displaystyle \underset{H}{|}}{C}}-\overset{\displaystyle \overset{H}{|}}{\underset{\displaystyle \underset{H}{|}}{C}}-H$$

This is often called *normal butane* and written *n*-butane. If we attach all the hydrogens to the branched butane C−C−C (commonly called isobutane) the cor-
                    |
                    C
responding formula would look like this:

_____ .

```
        H   H   H
        |   |   |
   H —  C — C — C — H
        |   |   |
        H   |   H
            |
        H — C — H
            |
            H
```

1-7   *Note:* These extended formulas showing which atoms are attached to each other, are called *structural formulas*. The simpler formulas, such as C_4H_{10}, which merely show the total number of atoms of each element in the molecule, are called *molecular formulas*.

Isobutane and *n*-butane are spoken of as *isomers* (from the Greek *isos*, equal, and *meros*, parts). Isomers are physically and/or chemically different substances with the same molecular formula.

The word *butane* is derived from the word *butter*. The unpleasant odor of rancid butter is due to an acid (butyric acid) containing four carbon atoms.

Now we have seen that there exist one methane (CH_4), one ethane (C_2H_6), and one propane (C_3H_8), but two substances with formula C_4H_{10}, whose carbon

"skeletons" were $C-C-C-C$ (*n*-butane) and
$C-C-C$ (isobutane) respectively.

The reason for the two butanes is the presence of
two different kinds of carbon atoms in propane
$(C-C-C)$: The propane molecule contains two
identical *end* carbon atoms each directly attached to
one carbon atom, and a middle carbon directly attached to *two* carbons.

We define a *primary* (1°) carbon as a carbon attached to *one* or *no* other carbon atoms; a *secondary*
(_____°) carbon as one attached to *two* other car-

bons, and a _____ (_____°) carbon as one
attached to three other carbons.

2°   tertiary (3°)

1-8   A quaternary (4°) carbon is one utilizing all
_____ of its bonds in attachments to other
(number)
_____ atoms.
(element)

---

four   carbon

1-9   Thus propane $(C-C-C)$ has _____ primary
carbon atoms and _____ secondary carbons.

---

2 primary   1 secondary: $(C-C-C)$
                              1° 2° 1°

1-10   If we attach a carbon atom to a primary carbon of propane, we obtain *n*-butane, $C-C-C-C$, which has _____ primary carbons and _____ secondary carbons.

---

2   2 $(C-C-C-C)$
        $1°$ $2°$ $2°$ $1°$

1-11   If, however, the fourth carbon is attached to the *secondary* carbon of propane, we obtain isobutane

$$C-\underset{\underset{C}{|}}{C}-C \quad \text{or} \quad \overset{C}{\underset{\underset{C}{|}}{\diagdown}}\overset{C}{\diagup}{C} \quad \text{or} \quad CH_3-\overset{CH_3}{\underset{CH_3}{\diagdown}}CH$$

where the center carbon is now a _____ ° carbon.

3° (or tertiary —attached to three other carbon atoms)

1-12   To write all possible pentanes (five-carbon alkanes) we must look at the two isomeric butanes and at the number of different positions in the butanes where a fifth carbon can be attached. There are two different kinds of carbons in *n*-butane ($C-C-C-C$). If a fifth carbon is attached at a primary carbon we obtain the skeleton formula _____.

$C-C-C-C-C$ (Note this may also be written:

$\underset{\underset{C}{|}}{C}-C-C-C$ or $C-\overset{\overset{C}{|}}{C}-C-C$, etc. The reason for

this will appear in Frame 1-17.)

1-13   If the fifth carbon, however, is attached to a secondary carbon of butane, the skeleton formula of the resulting pentane is _____.

$$C-\underset{\underset{C}{|}}{C}-C-C \quad \left(\text{or } C-C-\underset{\underset{C}{|}}{C}-C \right) \quad \text{This substance is}$$

called *isopentane*.

1-14   The second four-carbon molecule, isobutane

$$\left(\underset{\underset{CH_3}{\diagdown}}{\overset{\overset{CH_3}{\diagup}}{CH_3-CH}} \quad \text{or } C-C\underset{\diagdown C}{\overset{\diagup C}{}} \right) \quad \text{can be seen to have}$$

three identical CH_3 groups attached to a central tertiary carbon atom. A fifth carbon can therefore be attached either to an outer primary carbon (replacing an H of one of the CH_3 groups) or to the central carbon. The resulting carbon skeletons are

_____ and _____.

$$C-\underset{\underset{C}{|}}{C}-C-C \quad \text{and} \quad C-\underset{\underset{C}{|}}{\overset{\overset{C}{|}}{C}}-C$$

1-15   The first of these formulas, you will notice, is identical with the one in Frame _____. It is there named _____.

1-13   isopentane

1-16   The second formula in the answer to Frame 1-14

$$(C-\underset{\underset{\displaystyle C}{|}}{\overset{\overset{\displaystyle C}{|}}{C}}-C)$$ is a new pentane. Since the prefix *neo*

comes from the Greek *neos* meaning *new*, it was attached to the name pentane, so that the resulting name

for $C-\underset{\underset{\displaystyle C}{|}}{\overset{\overset{\displaystyle C}{|}}{C}}-C$ is _____ .

---

neopentane

1-17   No pentanes (of formula C_5H_{12}) with skeletons other
than the above three have been discovered. And no
structural formulas for molecules C_5H_{12} with different
skeletons can be devised. Note that the word skeleton is
an apt one for our formulas. Just as arms and legs can
move without destroying a person, so a skeleton formula can have its atomic positions shifted so long as
no bonds are broken. $C-\underset{\underset{\displaystyle C}{|}}{C}-C$ refers to the same

chemical as the formulas $\underset{\underset{\displaystyle C}{|}}{\overset{C\diagdown\quad\diagup C}{C}}$ and $\underset{\underset{\displaystyle C}{|}}{\overset{\overset{\displaystyle C}{|}}{C}}-C$. The

formulas $C-\underset{\underset{\displaystyle C}{|}}{C}-C-C$ and $C-C-\underset{\underset{\displaystyle C}{|}}{\overset{\overset{\displaystyle C}{|}}{C}}$ refer to (the

same/different) chemical(s).

---

the same chemical   (The second can be converted
into the first formula by turning it 180° in the plane
of the paper and bending one bond 90°. No bonds need
be broken to convert one formula into the other.)

1-18   Among the hexanes (6 carbons), neohexane is

$$C-\underset{\underset{C}{|}}{\overset{\overset{C}{|}}{C}}-C-C.$$ Neopentane and neohexane both contain

one _____ ° carbon atom. No other pentane
or hexane (or smaller molecule) contains anything
more complex than a tertiary (3°) carbon.

4° (quaternary)

1-19   There are more than three different hexanes,
C_6H_{14}. The number of different carbon skeletons for
this formula is _____ ; and the corresponding

skeleton structures are _____ , _____ ,

_____ , etc.

5; $C-C-C-C-C-C$ $C-\underset{\underset{C}{|}}{C}-C-C-C$

                            I   II

Names: *n*-hexane   isohexane

$C-C-\underset{\underset{C}{|}}{C}-C-C$ $C-\underset{\underset{C}{|}}{\overset{\overset{C}{|}}{C}}-C-C$ $C-\underset{\underset{C}{|}}{C}-\underset{\underset{C}{|}}{C}-C$

          III   IV   V

(not yet named)   neohexane   (not yet named)

1-20   If your answer to the previous frame was more than 5, you should convince yourself that some of your structures were duplicates—that is, they could be converted into each other simply by turning formulas and bending bonds. Note that ring structures such as C—C—C are excluded because they could not accomodate 14 hydrogen atoms.

C—C—C
| |
C—C—C

What shall we call Formulas III and V? We have run out of names and the invention of new prefixes is not very helpful for subsequently figuring out formulas. This is the reason chemists finally worked out a *systematic* naming system in which formulas could be derived from the information given in the name if a few simple rules were memorized. There are 9 heptanes (C_7H_{16}), 18 octanes (C_8H_{18}), 35 nonanes (C_9H_{20}), 75 decanes ($C_{10}H_{22}$), etc. You may want to convince yourself that some of these numbers are correct. We would need a lot of prefixes to distinguish them.

One final point before we leave hydrocarbons for a while: You may have noticed that all structures prefixed *iso* have as part of their molecule the three-carbon fragment,

C
  \
    C— , to which is attached, from
  /
C

the middle carbon, an unbranched chain of carbon atoms. Isoheptane (molecular formula C_7H_{16}) should

therefore have the skeleton structure ＿＿＿＿＿＿＿＿.

___

C
 \
  C—C—C—C—C
 /
C

PART B—THE COMMON NAMES OF SOME SIMPLE DERIVATIVES OF ALKANES

At this point you should be familiar with the meaning of primary, secondary, tertiary, and quaternary as designations for carbon atoms, and also with the prefixes *n-*, *iso-*, and *neo-*. We can now go on to discuss some "derivatives" of alkanes, that is, alkanes in which one or more of the hydrogen atoms have been replaced by atoms other than carbon or hydrogen.

1-21   One important class of compounds is the class of alkyl halides $C_nH_{2n+1}X$, where X is a halogen atom, F, Cl, Br, or I. If we write methane (CH_4) as CH_3-H, its chlorine derivative is written CH_3-Cl (or

$$H-\underset{\underset{H}{|}}{\overset{\overset{H}{|}}{C}}-Cl \text{ or } CH_3-Cl) \text{ and is called methyl chloride.}$$

Since C_2H_6 is ethane, C_2H_5Br, should, analogously, be

called _____ .

---

ethyl bromide

1-22   Just as we found in Part A that there were two different positions in propane (C_3H_8) to which a fourth carbon can be attached, so there are also two different propyl chlorides (C_3H_7Cl) that can be formed from propane by attachment of one chlorine atom to a carbon. One propyl chloride has the skeleton structure C—C—C—Cl. The skeleton struc-

ture of the other is _____ .

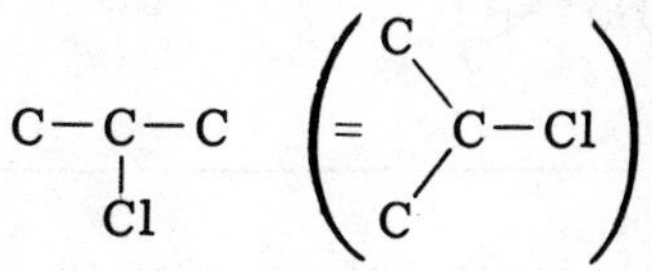

$$C-C-C \quad \left(= \begin{array}{c} C \\ \diagdown \\ \quad C-Cl \\ \diagup \\ C \end{array} \right)$$
$$\quad | $$
$$\quad Cl$$

1-23   How are these two propyl chlorides to be distinguished? The two butanes were prefixed

_____ and _____ .

---

n-   and   *iso-*

1-24   The same prefixes are used for the propyl chlorides so that similar structures as far as possible

can have similar names; $\begin{array}{c} C \\ \diagdown \\ \quad C-Cl \\ \diagup \\ C \end{array}$ is called

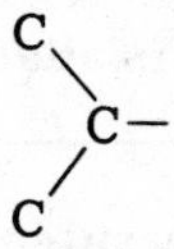

_____ propyl chloride; C—C—C—Cl is

called _____ propyl chloride.

---

isopropyl chloride   *n*-propyl chloride

1-25   Now isopropyl chloride has its chlorine atom attached to a (1°, 2°, 3°, 4°) carbon atom.
            (choose one)

2° (secondary)

$$\begin{pmatrix} C \\ \quad \boxed{C}\!-\!Cl \\ C \end{pmatrix}$$

1-26   Isopropyl chloride is, therefore, also (though
rarely) called secondary propyl chloride and written
sec-propyl chloride. The *four*-carbon alkyl chloride,
sec-butyl chloride, must have the structure

      _____ .

$$C\!-\!\boxed{C}\!-\!C\!-\!C \quad \begin{pmatrix} C-C \\ \quad \boxed{C}\!-\!Cl \\ C \end{pmatrix}$$
      |
     Cl

(Note that Cl is attached to the 2° carbon atom.)

1-27   Tertiary (written—and sometimes referred to as)

tert-butyl chloride has the structure _____ .

$$\begin{matrix} C \\ C\!-\!\boxed{C}\!-\!Cl \\ C \end{matrix} \quad \begin{pmatrix} C \\ | \\ = C\!-\!\boxed{C}\!-\!C \\ | \\ Cl \end{pmatrix}$$

1-28   *n*-butyl chloride should have the structure

         _____ , and isobutyl chloride the structure

         _____ .

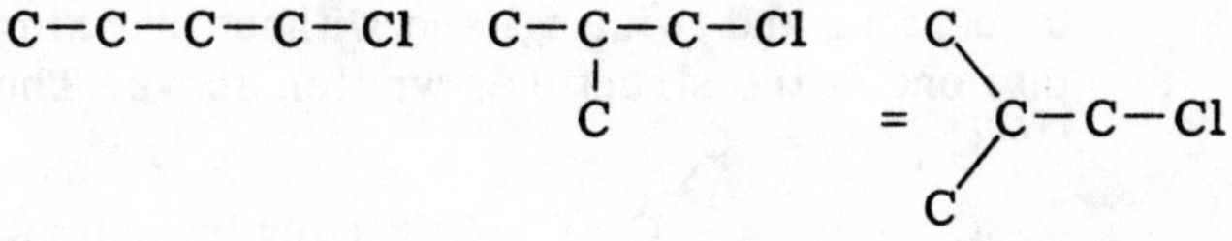

$$C-C-C-C-Cl \qquad C-\underset{\underset{C}{|}}{C}-C-Cl \qquad = \quad \underset{C}{\overset{C}{\diagdown}}C-C-Cl$$

     n-butyl chloride   isobutyl chloride
                                     (see Frame 1-20)

1-29   We have now distinguished four butyl chlorides by prefixes:

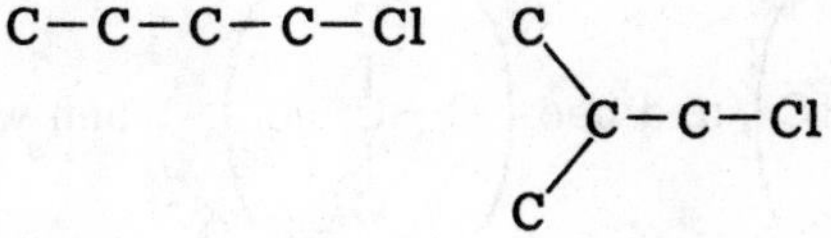

$$C-C-C-C-Cl \qquad\qquad \overset{C}{\underset{C}{\diagdown}}C-C-Cl$$

           n-   *iso*

$$\underset{C}{\overset{C-C}{\diagdown}}C-Cl \qquad\qquad \overset{C}{\underset{C}{C-}}C-Cl$$

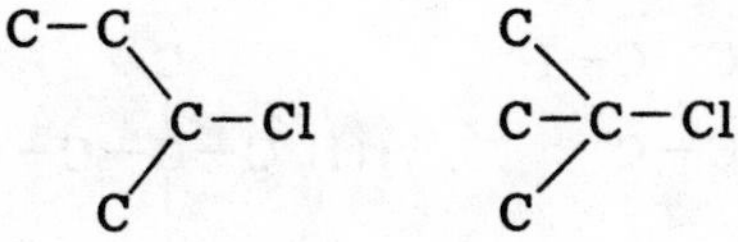

          sec-   *tert*-

Are there any further structures of formula C_4H_9Cl that cannot be converted, simply by twisting bonds, into one of these four structures? _____ .

No. If you have drawn further structures, you should convince yourself that each one is convertible simply by bending and twisting (and without breaking bonds) into one of the structures written above. Thus

$$C-C-C \quad = \quad \overset{\displaystyle C}{\underset{\displaystyle C}{\diagdown}} C-C-Cl = \text{isobutyl chloride}$$
$$\overset{|}{C}$$
$$\overset{|}{Cl}$$

1-30   We have found *two* butanes but *four* butyl chlorides (or bromides or iodides). In the five-carbon series, we found three pentanes: *n*- $(C-C-C-C-C)$, iso

$$\left(\overset{\displaystyle C}{\underset{\displaystyle C}{\diagdown}} C-C-C \right), \text{and neo} \left(C-\overset{\displaystyle C}{\underset{\displaystyle C}{\overset{|}{\underset{|}{C}}}}-C \right). \text{ When we attach}$$

one halogen atom, the resulting substances are known as pentyl or *amyl* iodides. There are eight such five-carbon iodides:

(a) $C-C-C-C-C-I$

(b) $C-\overset{|}{\underset{\displaystyle I}{C}}-C-C-C$

(f) $C-\overset{\displaystyle C}{\underset{\displaystyle I}{\overset{|}{\underset{|}{C}}}}-C-C$

(c) $C-C-\overset{|}{\underset{\displaystyle I}{C}}-C-C$

(d) $C-\overset{|}{\underset{\displaystyle C}{C}}-C-C-I$

(g) $C-\overset{\displaystyle C}{\underset{\displaystyle C}{\overset{|}{\underset{|}{C}}}}-C-I$

(e) $C-\overset{|}{\underset{\displaystyle I}{\overset{\displaystyle C}{\overset{|}{C}}}}-C-C$

(h) $C-C-\overset{\displaystyle C}{\overset{|}{C}}-C-I$

Which of the above formulas (a, b, c, d, e, f, g) should be named *n*-pentyl iodide ? _____ .

Which is *tert*-pentyl iodide ? _____ .

Which is isopentyl iodide (iso prefix always refers to formula $(CH_3)_2CH(CH_2)_nX$. $n = 0, 1, 2 \ldots$) ? _____ .

Which is neopentyl iodide ? _____ .

---

n-pentyl iodide = (a)   isopentyl iodide = (d)
tert-pentyl iodide = (f)   neopentyl iodide = (g)
Note: *n*-, *sec*-, and *tert*- are italicized and hyphenated when part of a name. Iso and neo are neither italicized not hyphenated.

1-31   Of the structures listed in the previous frame (a to g) three legitimately lay claim to the name *sec*-pentyl iodide because in each of them the iodine atom is attached to a secondary carbon atom. The three formulas are (list by letters a, b, c, etc.)

_____ , _____ , _____ .

(b) C—[C]—C—C—C, (c) C—C—[C]—C—C,
         | |
         I   I

         C
         |
(e) C—C—[C]—C
         |
         I

1-32   Since we do not want to distinguish the three *sec*-pentyl iodides by further prefixes, we have again come to a point where we should look for a new approach to nomenclature. Before we do so, however, we can generalize a little the naming system developed so far. The names for the alkyl halides can be used with slight modification for a large number of substances obtained by replacement of one hydrogen in an alkane by another atom or group of atoms. If R stands for an alkyl group (the part of the molecule containing only carbon and hydrogen) them $RONO_2$ is an alkyl nitrate (CH_3ONO_2 is methyl nitrate) and ROH is an alcohol (C—C—OH is ethyl alcohol just as C—C—Br is ethyl bromide). What then are the common names of

$$C-\overset{\underset{|}{OH}}{C}-C-C \ \text{ and } \ C-\overset{\overset{\displaystyle C}{|}}{\underset{\underset{\displaystyle C}{|}}{C}}-OH ? \ \underline{\hspace{3cm}},$$

_____.

$$C-\overset{\underset{|}{OH}}{C}-C-C$$

sec-butyl alcohol;

$$\overset{\displaystyle C}{\underset{\displaystyle C}{\overset{|}{C}}}-C-OH$$

tert-butyl alcohol

1-33

$$\overset{\displaystyle C}{\underset{\displaystyle C}{\overset{|}{C}}}-C-C-C-ONO_2 \text{ is named } \underline{\hspace{3cm}} \text{ and}$$

$$C-\overset{\overset{\displaystyle C}{|}}{\underset{\underset{\displaystyle C}{|}}{C}}-C-F \text{ is named } \underline{\hspace{3cm}}.$$

---

isopentyl nitrate   neopentyl fluoride

A Stop-Gap System of Nomenclature

PART A—THE SUBSTITUTED METHANE SYSTEM

2-1   Suppose the four hydrogen atoms of methane

$$H-\overset{\overset{\displaystyle H}{|}}{\underset{\underset{\displaystyle H}{|}}{C}}-H$$ are substituted by four *tert*-butyl groups

$(CH_3)_3C-$, to give the rather complex structure

$$\begin{array}{c}
C \quad C \quad C \\
\backslash | / \\
C \qquad C \qquad C \\
\backslash \quad | \quad / \\
C-C-C-C-C \\
/ \quad | \quad \backslash \\
C \qquad C \qquad C \\
/ | \backslash \\
C \quad C \quad C
\end{array}$$

that can be abbreviated $[(CH_3)_3C]_4C$. Our previous
naming system was quite unable to cope with such a
compound. But why not call it tetra-*tert*-butyl-
methane, a name that is to imply that the four hydrogen atoms of methane have been substituted by four
tert-butyl groups (tetra = 4 as in tetrahedron or
tetrarch)? If only two hydrogens have been replaced
by *tert*-butyl groups the name would be di-*tert*-butyl-
methane (di = 2 as in dialogue). Tetraethylmethane,
according to this system, represents a methane molecule with _____ hydrogen atoms of the methane
         (number)
replaced by ethyl groups.

4

2-2 If we write Et for ethyl, tetraethylmethane would

have the formula _____ .

Et₄C   or   Et—C—Et

with Et above and below the central C

$$Et_4C \quad or \quad Et-\underset{\underset{Et}{|}}{\overset{\overset{Et}{|}}{C}}-Et$$

2-3 Triethylmethane would only have _____

(number)

of the _____ hydrogen atoms of methane

(number)

replaced by _____ groups. One hydrogen

would therefore remain attached to the carbon.

---

3 of the 4;   ethyl

2-4 Triethylmethane, therefore, can be written

(ethyl = Et) as _____ .

$$Et_3CH \quad or \quad Et-\underset{\underset{H}{|}}{\overset{\overset{Et}{|}}{C}}-Et$$

2-5 Methylmethane is a methane with _____

(number)

hydrogen(s) replaced by methyl groups, while di-*n*-

propylmethane has _____ hydrogens of

(number)

methane replaced by _____ groups.

---

one   2 *n*-propyl groups

2-6   What if several different groups are attached to the carbon atom of what was once a methane molecule? In this program they are named in alphabetical order of alkyl groups (though they are sometimes listed according to size and complexity), each group named being attached separately to the central carbon atom that was once part of a methane molecule.

Therefore, diethylmethylmethane has _____
                                         (number)
methyl (Me) and _____ ethyl (Et) groups at-
                     (number)
tached to a carbon atom, while _____ hydro-
                                   (number)
gen atom(s) remain(s) unsubstituted by alkyl groups.

---

one (Me)   two (Et)   one (H)

2-7   Diethylmethylmethane, therefore, has the formula

_____ .

$$
\begin{array}{c}
\quad \text{Et} \\
\quad | \\
\text{Me}-\text{C}-\text{Et} \quad \text{or} \quad \text{Et}_2\text{CHMe} \\
\quad | \\
\quad \text{H}
\end{array}
$$

2-8   If we show all carbon and hydrogen atoms in a structural formula for diethylmethylmethane, we obtain the formula _____.

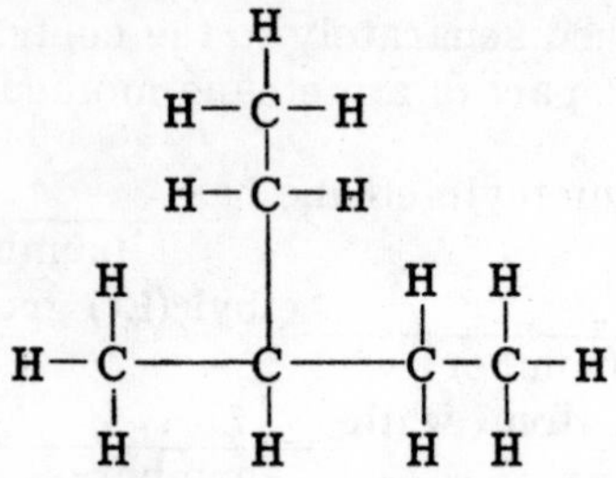

$$
\begin{array}{c}
H \\
| \\
H-C-H \\
| \\
H-C-H \\
| \\
\end{array}
$$

H—C——C——C—C—H

(structure shown with all H atoms)

2-9   The name of the substance whose structural formula can be represented as

$$
\left(
\begin{array}{l}
di = 2 \\
tri = 3, \text{ as in} \\
\quad \text{tricycle} \\
tetra = 4
\end{array}
\right)
$$

C
|
C
|
C—C—C—C—C
|
C

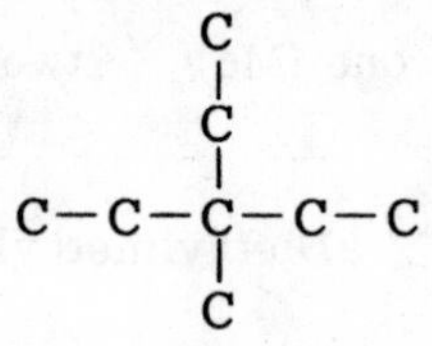

is _____.

---

triethylmethylmethane   (Note that the name is written as a single word. Since methane is CH_4, it would be confusing to write it as a separate word in a larger name.)

2-10 Me_2CEt_2 may be named _____.

---

diethyldimethylmethane

2-11   When we come to propyl groups, isopropyl is named before *n*-propyl. Among the butyl groups *n*-, *sec*-, and *tert*- are named before iso. The procedure is to use alphabetical order of groups, but ignoring italicized prefixes until the main order is established, thus: *n* butyl, isobutyl, isopropyl, *n*-propyl. Therefore, *n*-butylethylisobutyl-*n*-propylmethane has the formula

_____ (use *n*-Bu, isobu, Et, *n*-Pr).

$$
\begin{array}{c}
\text{isobu} \\
| \\
n\text{-Bu}-\text{C}-\text{Et} \\
| \\
n\text{-Pr}
\end{array}
$$

2-12   In a complex formula to be named by this system, the carbon to be chosen as the one that was once a methane carbon is the most substituted carbon (i.e., a quaternary rather than a tertiary, and so on).

$$
\begin{array}{c}
\text{C} \\
| \\
\text{C}-\text{C}-\text{C}-\text{C}-\text{C} \\
|| \\
\text{C}\text{C}
\end{array}
$$ is named *sec*-butyltrimethylmethane

rather than *tert*-butylethylmethane. The former name lists the groups attached to a quaternary carbon (fourth from the left in the chain), while the latter lists the groups around a tertiary carbon (middle of chain). By the substituted methane system, the formula

$$
\begin{array}{c}
\text{C}-\text{C}-\text{C} \\
| \\
\text{C}-\text{C}-\text{C}-\text{C}-\text{C}-\text{C}-\text{C} \\
| \\
\text{C}-\text{C}-\text{C}
\end{array}
$$

must be named _____.

---

tetra-*n*-propylmethane

2-13

$$C-C-C-\overset{\overset{\displaystyle C-C}{|}}{C}-C-\overset{\overset{\displaystyle C}{|}}{C}-C \quad \text{must be named}$$

_____.

---

ethylisobutylmethyl-*n*-propylmethane

But what shall we do with the formula

$$C-C-\overset{\overset{\displaystyle C}{|}}{\underset{\underset{\displaystyle C}{|}}{C}}-C-C-C-\overset{\overset{\displaystyle C}{|}}{\underset{\underset{\displaystyle C}{|}}{C}}-C \quad \text{or even the simpler formula}$$

$$C-C-C-C-C-\overset{\overset{\displaystyle C}{|}}{\underset{\underset{\displaystyle C}{|}}{C}}-C? \quad \text{They have no name by either of}$$

the two systems we have discussed. The substituted methane system works for a few more compounds than the common naming system, but when the substituent groups become so complex that they have no name in the common naming system, the new system fails us too.

PART B—ANOTHER STOP GAP SYSTEM: THE CARBINOL SYSTEM FOR ALCOHOLS

2-14   Alcohols contain OH groups. The OH group is attached to a carbon atom which in turn is attached to three other atoms, either H or C. The carbon atom together with the OH group is called the carbinol group. Attached groups, other than hydrogen, are named in alphabetical order as before but instead of adding the word methane we add the word carbinol. If less than three groups are named, the remaining valences are attached to hydrogen atoms. Triethyl-carbinol must have the formula _____ .

Et$_3$COH [or (CH$_3$CH$_2$)$_3$COH   or   Et—C—Et   or

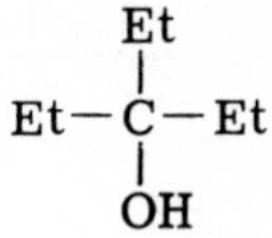

$$
\begin{array}{c}
\text{Et} \\
| \\
\text{Et—C—Et} \\
| \\
\text{OH}
\end{array}
$$

$$
\begin{array}{c}
\text{C} \\
| \\
\text{C} \\
| \\
\text{C—C—C—C—C} \\
| \\
\text{OH}
\end{array}
$$
 ]

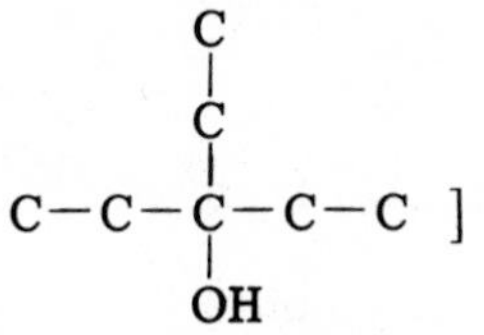

2-15   *n*-butyl-*n*-pentyl-*n*-propylcarbinol has the formula

_____ .

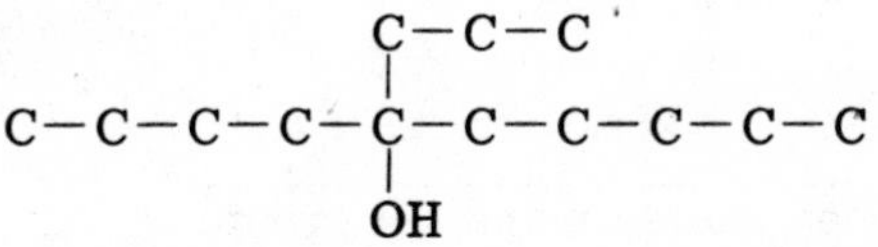

$$
\begin{array}{c}
\text{C—C—C} \\
| \\
\text{C—C—C—C—C—C—C—C—C—C} \\
| \\
\text{OH}
\end{array}
$$

2-16   methylcarbinol is _____ .
                    (formula)

CH$_3$CH$_2$OH (the same as ethyl alcohol).   Note that only one valence is used by a methyl group here. The remaining valences are occupied by hydrogens.

2-17   Carbinol itself must, therefore, have the formula

_____.

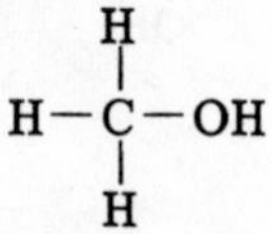

$$H-\underset{\underset{H}{|}}{\overset{\overset{H}{|}}{C}}-OH$$

2-18   The common name for carbinol is _____.

---

methyl alcohol

2-19   Tri-*tert*-butylcarbinol has the formula

_____.
(show carbon skeleton and OH group)

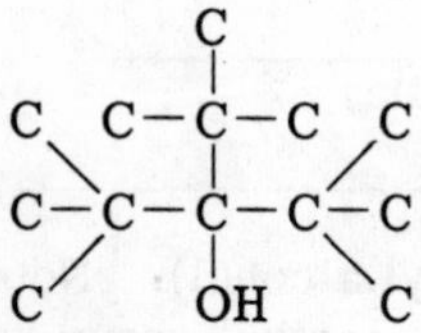

2-20 According to the carbinol system, the name for

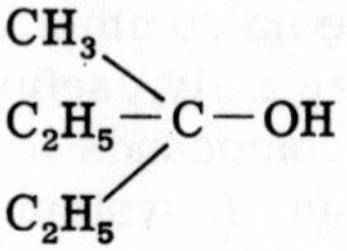

$$
\begin{array}{c}
CH_3 \\
C_2H_5-C-OH \quad \text{is} \underline{\hspace{3cm}}. \\
C_2H_5
\end{array}
$$

---

diethylmethylcarbinol

2-21

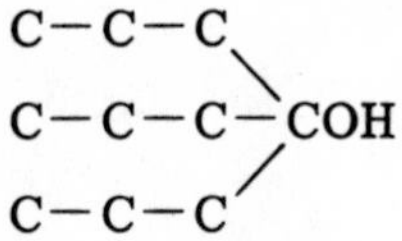

$$
\begin{array}{c}
C-C-C \\
C-C-C-COH \quad \text{is named} \underline{\hspace{3cm}}. \\
C-C-C
\end{array}
$$

---

tri-*n*-propylcarbinol

2-22

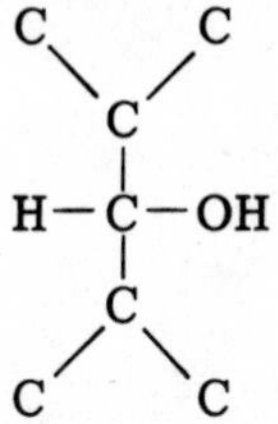

$$
\begin{array}{c}
C \qquad C \\
C \\
H-C-OH \quad \text{is named} \underline{\hspace{3cm}}. \\
C \\
C \qquad C
\end{array}
$$

---

diisopropylcarbinol

2-23

$$
CH_3 \\
\text{And } CH_3CH_2CH_2CH_2-C-CH_3 \text{ is named} \underline{\hspace{2cm}}. \\
OH
$$

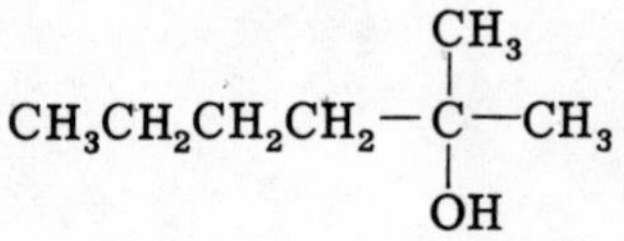

---

n-butyldimethylcarbinol

Again, our system breaks down as soon as groups that are attached to the carbinol carbon have no common names. Though these stop-gap names are occasionally useful, and have become common names for a few compounds (such as triethylcarbinol), they are not widely used. A system needed to be developed that was not limited simply by the size of a molecule or the number of its branches. The obvious procedure was to use numbers rather than prefixes as far as possible. This was the procedure adopted at a meeting of the International Union of Pure and Applied Chemistry. The system is described in Chapter 4.

The Common Names of Amines

Amines are alkyl derivatives of ammonia, NH_3. The rules for arriving at their common names are so simple that they will be stated at the beginning and then a number of examples given to convince you that you understand them.

One, two, or three hydrogens of ammonia can be replaced by alkyl groups. The groups are named in alphabetical order and the word *amine* is attached. The system is quite analogous to the substituted methane and the carbinol system, and we again use the prefixes di-, tri-, and tetra- for multiple groups of the same kind.

3-1   If one alkyl group (R) is mentioned, the general formula will be RNH_2; if two groups (R, R', which may be the same or different), the general formula will be R—N—H, and if three groups (R, R', and R'')
  |
  R'
the formula will be _____ .

R—N—R'
|
R''

3-2   The formula for ethylamine must be _____ .

$C_2H_5NH_2$ or

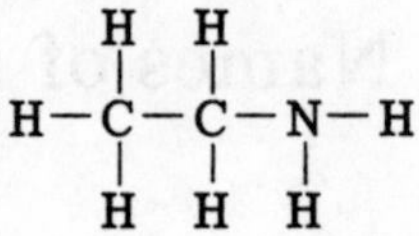

$$H-\underset{\underset{H}{|}}{\overset{\overset{H}{|}}{C}}-\underset{\underset{H}{|}}{\overset{\overset{H}{|}}{C}}-\underset{\underset{H}{}}{\overset{}{N}}-H$$

3-3   Dimethylamine is _____ .

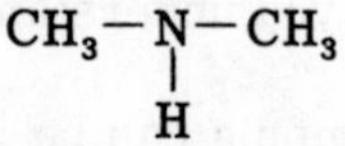

$$CH_3-\underset{\underset{H}{|}}{N}-CH_3 \;=\; (CH_3)_2NH$$

3-4 $C_2H_5-\underset{\underset{CH_3}{|}}{N}-H$ is named _____ .

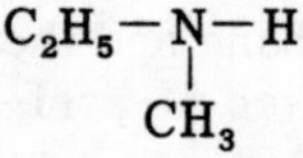

ethylmethylamine   (Note that *amine* never appears
as a separate word in a name.)

3-5   The skeleton structure for triisopropylamine is

_____ .

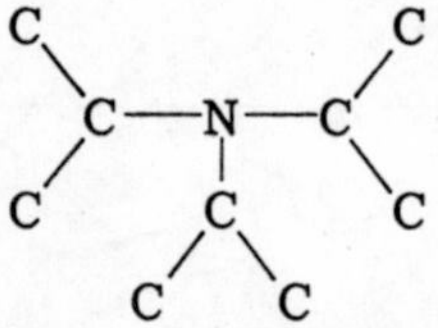

$$\underset{\underset{C}{|}}{\overset{\overset{C}{|}}{C}}-N-\underset{\underset{C}{|}}{\overset{\overset{C}{|}}{C}}$$

3-6   Isopropylmethyl-*n*-propylamine has the structure

_____.

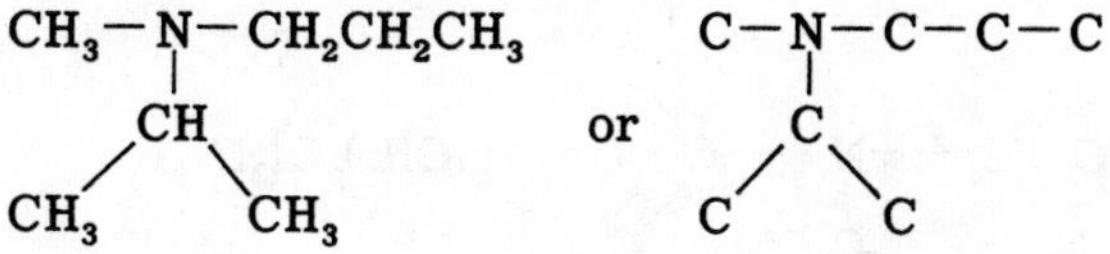

$$CH_3-N-CH_2CH_2CH_3 \qquad\qquad C-N-C-C-C$$

CH₃—N—CH₂CH₂CH₃   C—N—C—C—C
      | |
     CH   or   C
   /   \\   / \\
CH₃   CH₃   C   C

3-7   The name for $CH_3NHC_2H_5$ would be _____.

ethylmethylamine

3-8   And the name for $(CH_3)_3N$ _____.

trimethylamine

3-9   What is the formula for *tert*-butylamine (don't confuse with tributylamine)?

_____

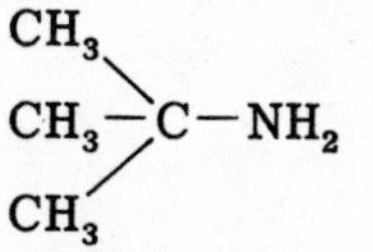

$$CH_3$$
$$CH_3-C-NH_2 \quad or \quad (CH_3)_3CNH_2$$
$$CH_3$$

Note that only one H of ammonia is substituted.

3-10   What would be the skeleton structure of tri-*tert*-

butylamine ? _____

$$\begin{array}{ccccc} & \text{C} & & & \text{C} \\ & | & & & | \\ \text{C}-\text{C} & - & \text{N} & - & \text{C}-\text{C} \quad = [(CH_3)_3C]_3N \\ & | & | & & | \\ & \text{C} & \text{C} & & \text{C} \\ & & / | \backslash & & \\ & & \text{C}\ \text{C}\ \text{C} & & \end{array}$$

4

The Systematic Naming of Alkanes

In working through the common names of hydrocarbons we ran out of prefixes after defining the meanings of *n*-, iso, *sec*-, *tert*- and neo. Now there is no difficulty in thinking up new prefixes; the problem is to remember their meanings. The situation finally became so confused that an international congress was called at Geneva, Switzerland in 1892 to devise a naming system by which not only presently known compounds but also those to be prepared in the future could be given names understandable to anyone trained in the system. A "Definitive Report of the Commission on the Reform of the Nomenclature of Organic Chemistry" was unanimously adopted by the International Union of Chemistry in 1930 and was extended in 1936, 1938, and 1957. The international organization, now known as the International Union of Pure and Applied Chemistry, is abbreviated IUPAC. Names based on the international rules are called Systematic names, Geneva names or IUPAC names. The rules, with comments, were published in *J. Am. Chem. Soc.* **82**, 5545-5584 (1960). The very closely related set of rules for indexing compounds in *Chemical Abstracts* was published as the introduction to Volume 56 of the Subject Index of *Chemical Abstracts* (1962) (see Bibliography, p. 193).

4-1   The first rule of the IUPAC system is to look for the longest chain of carbon atoms, eliminating all branches off the longest continuous chain. This chain is known as the "parent chain." Branches are named subsequently.

The longest continuous chain in isobutane
C — C — C is a chain of _____ carbon atoms.
$\quad\quad$ | (number)
$\quad\quad$ C

---

3 $\quad$ 1 $\quad$ 2 $\quad$ 3
(C — C — C)
$\quad\quad$ |
$\quad\quad$ C

4-2 $\quad$ Isobutane is therefore considered a derivative of the "parent" three-carbon alkane whose molecular formula is _____; this was called p____e.

C_3H_8 $\quad\quad$ propane

4-3 $\quad$ Before we proceed we should review some earlier information needed in systematic naming:
The two-carbon alkane is called _____.

ethane

4-4 $\quad$ The simplest alkane, CH_4, is called _____.

methane

4-5 $\quad$ To convert propane, C_3H_____ , to isobutane, C_4H_____ , we need to *replace one hydrogen atom* of propane by a group C_____ H_____

$C_3H_{\underline{8}}$ $\quad\quad$ $C_4H_{\underline{10}}$ $\quad\quad$ CH_3 $\quad$ or C_1H_3 (methyl)

4-6   If instead we had replaced a hydrogen atom in propane, C_3H_8, by a *two*-carbon group, in order to form a new alkane, C_5H_{12}, that *group* would have had to have the formula _____.

C_2H_5

4-7   A one-carbon group that can replace a hydrogen atom has the formula CH_3; and a two-carbon group, C_2H_5. An n-carbon group that can replace a hydrogen atom should then have the formula C_nH_____.

$2n + 1$ i.e., C_nH_{2n+1}

4-8   We say that the group CH_3 is derived from CH_4 by removing one hydrogen atom; since CH_4 is methane, CH_3 is called methyl. Since C_2H_6 is ethane, C_2H_5 is called _____.

---

ethyl (In general, we replace the ending -ane by -yl.)

4-9   Methyl, ethyl, propyl, etc. are called *alkyl* groups. Thus, C_4H_9 is called _____ and is an example of an _____ group.

---

butyl   alkyl

4-10 C_3H_8 is _____; and C_3H_7 is _____.

---

propane   propyl

4-11   Now we can return to the systematic naming of
isobutane. Its molecular formula is C_4H_{10}. Its carbon
"skeleton" looks like this: C—C—C. Its longest un-
                                          |
                                          C

branched carbon chain contains three carbon atoms.
The name of the three-carbon "parent" alkane is

_____.

---

propane

4-12   In addition to the three-carbon chain, isobutane
contains a one-carbon group attached to the middle
carbon. This one-carbon *group* has the formula

_____ and is called a _____ group.

CH_3 methyl

4-13   Isobutane, therefore, has the IUPAC name methyl-
propane. More precisely it is named 2-methylpro-
pane. Except in the simplest cases, the point of attachment of any branch must be specified. Thus there
are two methylpentanes:

```
1   2   3   4   5        5   4   3   2   1        1   2   3   4   5
C — C — C — C — C   =   C — C — C — C — C        C — C — C — C — C
    |                               |                        |
    C                               C                        C
      2-methylpentane                       3-methylpentane
```

We number the longest or "parent" chain, beginning at
the end that makes the point of attachment occur at
the smallest possible number. Since a six-carbon hydrocarbon is named *hexane* (prefixes such as *n*- are

not used in the IUPAC system), the Systematic name

for C—C—C—Ċ—C—C must be _____ .

with C above the 4th carbon (Ċ shown as C—bonded above)

3-methylhexane

$$C—C—C—\overset{\overset{\textstyle C}{|}}{C}—C—C$$
6 5 4 3 2 1

4-14   The compound with the structural formula

$$H—C—C—C—C—C—C—C—H$$ (each carbon with H above and below) is *heptane*. To simplify formula writing, the heptane formula is often written simply as $CH_3CH_2CH_2CH_2CH_2CH_2CH_3$ or even as $CH_3(CH_2)_5CH_3$. Such formulas are spoken of as "condensed formulas." Similarly the formula of pentane (five carbons) may be written in condensed form as

_____ or _____ .

$CH_3CH_2CH_2CH_2CH_3$ or $CH_3(CH_2)_3CH_3$

4-15   Note that a five-carbon chain need not be written on a horizontal line. The following carbon skeletons all represent *pentane*

C—C—C—C—C   C—C—C—C with C below   C—C—C with C—C below

C—C / C—C—C   C—C / C—C / C

Remember that two different formulas represent the same chemical compound if one formula can be con-

verted into the other simply by twisting the formula
but without breaking any bonds. By contrast,
C—C—C—C—C can only be converted into
C—C—C—C if at least one C—C bond and one C—H
　　　|
　　　C

bond are broken. Thus C—C—C—C is a methylbutane
　　　　　　　　　　　　　　|
　　　　　　　　　　　　　　C

because the longest continuous chain contains four
　　　　　　　　　　　　　　　C
　　　　　　　　　　　　　　　|
carbon atoms. On the other hand, C—C—C—C is
　　　　　　　　　　　　　　　　　　　　|
　　　　　　　　　　　　　　　　　　　　C

hexane. C—C—C　　is _____ .
　　　　　|　　　|
　　　　　C　　　C—C
　　　　　|
　　　　　C

---

heptane

4-16　　The "parent" chain or longest continuous chain of
carbon atoms (not necessarily in a straight line) in
the carbon skeleton C—C—C—C—C contains
　　　　　　　　　　　　　　　　　　|
　　　　　　　　　　　　　　　　　　C
　　　　　　　　　　　　　　　　　　|
　　　　　　　　　　　　　　　　C—C—C

_____ carbon atoms.
(number)

　　　　　　1　2　3
　7　　　C—C—C—C—C
　　　　　　　　|
　　　　　　　　C4
　　　　　　　　|
　　　　　　　C—C—C
　　　　　　　5　6　7

4-17   For convenience in naming, let us rewrite our

               1   2   3

skeleton formula, C—C—C—C—C, in such a way as

                         |

                       C4

                       |

                  C—C—C

                  5   6   7

to put the parent chain of seven carbon atoms in a

                  1   2   3   4   5   6   7

straight line, thus C—C—C—C—C—C—C

                            |

                            C

                            |

                            C

You should make quite sure that this can be done without the breaking of any bonds. The molecule is now
seen to be a substituted heptane. To the parent hep-

tane chain is attached one branch of _____ carbon
atoms.

---

2 (C_2H_5, ethyl)

4-18   C—C—C—C—C—C—C   Numbering the parent

                   | chain from that end

                   C   which makes the branch

                   | number the smallest pos-

                   C   sible, leads to position

                                       number _____

                                       for the branch.

---

         1   2   3   4   5   6   7

3   C—C—C—C—C—C—C

                  |

                  C

                  |

                  C

4-19   C—C—C—C—C—C—C   or   C—C—C—C—C is
$\qquad\qquad$|$\qquad\qquad\qquad\qquad\qquad\quad$|
$\qquad\qquad$C$\qquad\qquad\qquad\qquad\qquad\quad$C
$\qquad\qquad$|$\qquad\qquad\qquad\qquad\qquad\quad$|
$\qquad\qquad$C$\qquad\qquad\qquad\qquad\qquad$C—C—C

3-ethylheptane. The condensed formula $(CH_3CH_2)_3CH$

can be expanded to

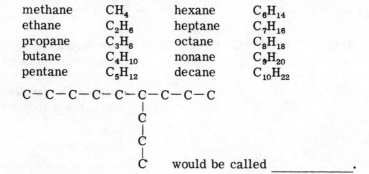

must be _____.

3-ethylpentane

4-20   The first ten unbranched alkanes have the IUPAC
names and formulas as follows:

methane	CH_4	hexane	C_6H_{14}
ethane	C_2H_6	heptane	C_7H_{16}
propane	C_3H_8	octane	C_8H_{18}
butane	C_4H_{10}	nonane	C_9H_{20}
pentane	C_5H_{12}	decane	$C_{10}H_{22}$

C—C—C—C—C—C—C—C—C
$\qquad\qquad\qquad\qquad\quad$|
$\qquad\qquad\qquad\qquad\quad$C
$\qquad\qquad\qquad\qquad\quad$|
$\qquad\qquad\qquad\qquad\quad$C
$\qquad\qquad\qquad\qquad\quad$|
$\qquad\qquad\qquad\qquad\quad$C$\qquad$would be called _____.

4-propylnonane   (Note:The prefixes *penta* to *deca* are
derived from the Greek or Latin for the numbers 5 to
10: thus *penta*gon, *hexa*meter, *hepta*teuch, *octa*ve,
*nona*genarian, and *deca*logue.)

4-21   The formula

$$C—C—\underset{\underset{C}{|}}{C}—C—C—\underset{\underset{C}{|}}{C}—C—C—C—C$$

is named 3, 6-dimethyldecane. 3,6-dimethyl designates that there are *two* methyl groups attached to the dec-ane chain, one at the third and one at the sixth carbon atom. If both methyls were attached at the same carbon atom of the parent chain, *the number designating that carbon would be repeated*: 2,2-dimethylpropane has the

carbon skeleton _____ .

$$C—\underset{\underset{C}{|}}{\overset{\overset{C}{|}}{C}}—C$$

4-22   Three methyl branches are written as trimethyl; four ethyl branches, tetraethyl. 3,3,5-triethyloctane

has the carbon skeleton _____ .

$$C—C—\underset{\underset{\underset{C}{|}}{\overset{C}{|}}}{\overset{\overset{\overset{C}{|}}{C}}{C}}—C—\underset{\underset{\underset{C}{|}}{\overset{C}{|}}}{C}—C—C—C$$

4-23   If several groups appear as branches, they are
usually listed in alphabetical order of alkyl groups
and it is this practice we shall follow. Thus,
C—C—C—C—C—C—C—C—C—C is 3-ethyl-5-

```
            C       C       C
            |       |       |
            C       C
                    |
                    C
```

methyl-7-propyldecane. The carbon skeleton for

5-butyl-4-methylnonane is _____.

```
C—C—C—C—C—C—C—C—C
      |   |
      C   C—C—C—C
```

4-24   What is the name of

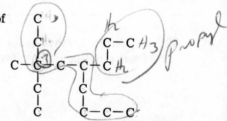

_____? (Remember to locate the
longest continuous chain first—and that it does not
have to be in a straight line in the formula. Name substituent groups in alphabetical order. Remember, too,
that numbering is started from that end which gives the
first branch encountered the smaller number, regardless of the size of the branch.)

3-ethyl-3-methyl-5-propylnonane

```
              C
              |
              C        C—C
              |   4  5 |
         C—³C—C—C—C
          3 |        |
          ²C          C6
           |          |
           C          C—C—C
           1          7  8  9
```

4-25   Name the following: (a) $CH_3(CH_2)_7CH_3$ _____

(b) $(CH_3CH_2)_4C$ _____

(Write out the expanded skeletons for these formulas before naming unless you are quite sure what they represent.)

(a) nonane   (b) 3,3-diethylpentane   (Note commas are used to separate numbers; hyphens to join numbers and words.)

4-26   2-methyl-4,5-dipropyloctane has the skeleton

structure _____.

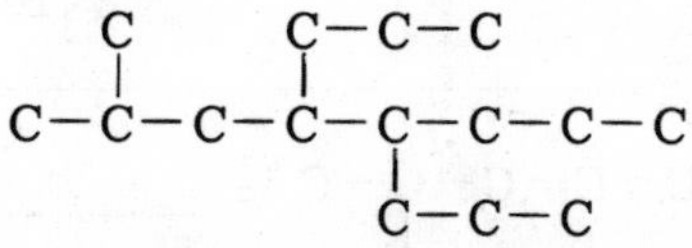

```
        C        C—C—C
        |        |
  C—C—C—C—C—C—C—C
                 |
                 C—C—C
```

Note that *methyl* precedes *dipropyl* in the name, because only the names of alkyl groups are considered for alphabetical ordering. Prefixes di-, tri-, tetra- are attached *after* the order of the groups is established.

4-27   What is the IUPAC name for

```
                C¹⁰
           C    C⁵
           |    |₃
   C—C—C⁴—C—C²—C—C¹
           |
           C⁵
           |
           Cᵃ
           |
      ₃ C—C̆—C¹
```
_____.

3,5-diethyl-5-methyldecane

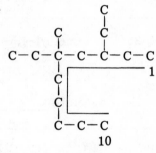

4-28   The name for   C—C—C—C—C is _____
                                  |
                                  C—C

(careful!)

3-methylhexane

4-29   The skeleton structure of 3,3,5-triethylheptane is

_____.

```
         C—C   C—C
         |     |
   C—C—C—C—C—C—C
         |
         C—C
```

The Systematic Naming of Alkenes, Alkynes, Dienes, and Simple Cyclic Hydrocarbons

5-1   The molecules we have discussed so far have pairs of atoms linked by single bonds only. Ethane, for instance, has the structural formula

$$
\begin{array}{c}
\ \ \ \ \text{H}\ \ \ \text{H} \\
\ \ \ \ |\ \ \ \ \ | \\
\text{H}-\text{C}-\text{C}-\text{H} \\
\ \ \ \ |\ \ \ \ \ | \\
\ \ \ \ \text{H}\ \ \ \text{H}
\end{array}
$$

Suppose, however, that two carbon atoms were linked to each other by a pair of bonds, thus $C=C$. Then, since carbon has a valence of four, only two other atoms can be linked by each carbon. The simplest hydrocarbon containing a double bond, therefore, has the molecular formula C_2H_4 and the structural formula _____.

$$
\begin{array}{c}
\text{H}\ \ \ \text{H} \\
|\ \ \ \ \ | \\
\text{C}=\text{C} \\
|\ \ \ \ \ | \\
\text{H}\ \ \ \text{H}
\end{array}
\quad \text{or} \quad
\begin{array}{c}
\text{H}\ \ \ \ \ \ \ \ \ \text{H} \\
\ \diagdown\ \ \ \ \ \diagup \\
\ \ \ \text{C}=\text{C} \\
\ \diagup\ \ \ \ \ \diagdown \\
\text{H}\ \ \ \ \ \ \ \ \ \text{H}
\end{array}
\quad \text{or, in condensed form,}
$$

$H_2C=CH_2$ or $CH_2=CH_2$

5-2   The substance of molecular formula C_2H_4 is commonly known as ethylene. Its systematic name is *ethene*. Hydrocarbons containing one double bond are known as *alkenes*. Alk*enes* may be thought of as being derived from alk*anes* by the removal of a pair of hydrogen atoms from adjacent carbons. These adjacent carbons become doubly linked. In naming an alkene, we first name the alkane with the same number of carbon atoms and then replace the ending *ane* by *ene*. Since C_3H_8 is named propane, the name of the compound with the formula C_3H_6 should be _____.

---

propene   (Its condensed structural formula may be written $CH_3-CH=CH_2$)

5-3   Since alkanes have the general formula C_nH_{2n+2}, and alkenes are derived from them by the loss of two hydrogen atoms, the general formula for alkenes is

_____.

C_nH_{2n} (Note: This is also the formula for cycloalkanes $(CH_2)_n$, discussed in Frame 5-22.)

5-4   The molecular formula for ethene is C_2H_4. For the three-carbon alkene, propene, it is C_3H_6. For a five-carbon alkene the molecular formula must be

_____.

C_5H_{10}

5-5   In converting butane, $CH_3CH_2CH_2CH_3$, to an alk*ene*, two different structures can be formed. The pair of hydrogens may be lost either from the two middle carbons or from a middle and the adjacent end carbons. The resulting butenes have the formulas and names:

$$H-\overset{\displaystyle H}{\underset{\displaystyle H}{C}}-\overset{\displaystyle H}{C}=\overset{\displaystyle H}{C}-\overset{\displaystyle H}{\underset{\displaystyle H}{C}}-H \quad \text{and}$$

2-butene

$$H-\overset{\displaystyle H}{C}=\overset{\displaystyle H}{C}-\overset{\displaystyle H}{\underset{\displaystyle H}{C}}-\overset{\displaystyle H}{\underset{\displaystyle H}{C}}-H \;=\; H-\overset{\displaystyle H}{\underset{\displaystyle H}{C}}-\overset{\displaystyle H}{\underset{\displaystyle H}{C}}-\overset{\displaystyle H}{C}=\overset{\displaystyle H}{C}-H$$

1-butene

The number of different positions for a double bond in a five-carbon chain must be _____.

---

2   (i.e., $C=C-C-C-C$ and $C-C=C-C-C$)

5-6   The number of different positions for a double bond in a six-carbon chain must be _____.

---

3   ($C=C-C-C-C-C$; $C-C=C-C-C-C$;
    $C-C-C=C-C-C$)

5-7   In Frame 5-5 we gave the name of $CH_2=CH-CH_2-CH_3$ as 1-butene and of $CH_3-CH=CH-CH_3$ as 2-butene. The IUPAC rule is to number the longest continuous chain of carbons that contains the double bond from the end that places the double bond at the lowest numbered carbons. Accord-

ing to this rule the double bond in the hexene
$CH_3-CH_2-CH_2-CH=CH-CH_3$ occurs between

carbon atoms _____ and _____.

---

2   3

5-8   Since double bonds always occur between adjacent
carbons, it is superfluous to indicate the numbers of
both carbons terminating the double bond. It is enough
to indicate the lower number. Thus, 2-hexene has the
formula $CH_3-CH=CH-CH_2-CH_2-CH_3$. The formula
for 3-hexene, written in the same form (a condensed

structural formula) would be _____.

$CH_3-CH_2-CH=CH-CH_2-CH_3$

5-9   The name for $CH_3-CH=CH-CH_2-CH_2-CH_2-CH_3$

is _____. (C_5 pent, C_6 hex, C_7 hept, C_8 oct)

2-heptene

5-10   When branches occur, we look for the longest
carbon chain *of which the double bond is a part* (this
is the "parent" alkene here). Thus
$CH_3-CH_2-CH_2-\underset{\underset{CH_2}{\|}}{C}-CH_2-CH_3$ will be named as a

substituted *pentene*, not a substituted *hexane*. Its
name is 2-ethyl-1-pentene. This may be seen more
easily by rewriting the formula as
$CH_2=\underset{\underset{CH_2-CH_3}{|}}{C}-CH_2-CH_2-CH_3$

The parent alkene in the formula $CH_2{=}C{-}CH_3$
                                        |
                                  $CH_2{-}CH_2{-}CH_3$

is _____ . (Give name including numbered
position of double bond.)

1-pentene

5-11   The systematic name of $CH_2{=}C{-}CH_3$
                                        |
                                  $CH_2{-}CH_2{-}CH_3$

is therefore _____ .

2-methyl-1-pentene

5-12   Note that the numbering is determined by the location of the double bond only, and not by the position of
the branches. Thus $(CH_3)_2CHCH_2CH_2CH_2CH_2CH{=}CH_2$
is 7-methyl-1-octene (*not* 2-methyl-7-octene).
     Accordingly, $(CH_3)_2CH{-}CH{=}CH{-}CH_3$ is named

_____ .

4-methyl-2-pentene

5-13   The name of the alkene whose carbon skeleton is
$C{-}C{-}C{-}C{-}C{=}C{-}C$
              | |
              C   C
              |
              C   must be _____ .

4-ethyl-2-methyl-2-heptene

5-14   The name for   C—C is _____ .

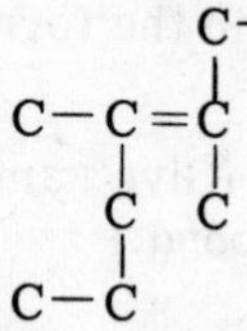

$$C-C=C$$
$$\quad\quad | \quad |$$
$$\quad\quad C \quad C$$
$$C-C$$

3,4-dimethyl-3-heptene

5-15   The carbon skeleton of the substance 2-methyl-6-

propyl-4-nonene may be written _____ .

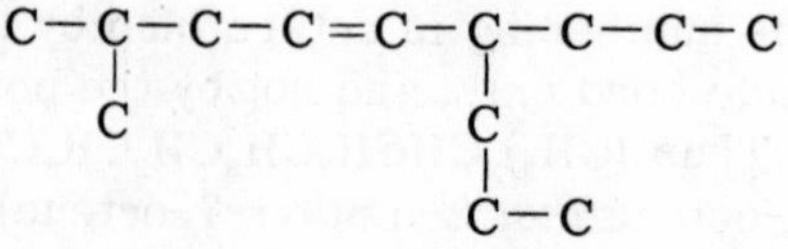

C—C—C—C=C—C—C—C—C
      | |
      C   C
                     |
                     C—C

5-16   *Polyenes—Hydrocarbons with two or more double bonds*

$CH_2=CH-CH=CH_2$ is named 1,3-butadiene.

The ending for a molecule with three double bonds is -atriene, with four -atetraene, and so on.

$CH_2=CH-CH=CH-CH=CH_2$ is accordingly

named _____ .

1,3,5-hexatriene

5-17   In branched hydrocarbon polyenes, the longest carbon chain containing the double bonds is numbered and named first; the branches are then numbered and named, and written preceding the parent name.

"Isoprene," the "building block" of natural rubber has the formula

$$\overset{\displaystyle CH_3}{\underset{\displaystyle |}{CH_2=C-CH=CH_2}}$$

Its systematic name is 2-methyl-1,3-butadiene. The formula for 2,3-dimethyl-2,4,5-octatriene is

_____. (Show in a similar form to the formula above, and remember that the valence of carbon is 4.)

$$CH_3-\overset{\displaystyle CH_3}{\underset{\diagup}{C}}=\overset{\displaystyle CH_3}{\underset{\diagup}{C}}-CH=C=CH-CH_2-CH_3$$

5-18   *Alkynes* (pronounced to rhyme with *lines*). An alk*yne* contains a triple bond C≡C. *Polyynes* contain two or more such bonds. The rules for naming follow those described for alkenes. The simplest alkyne has the molecular formula _____.

C_2H_2

5-19   The systematic name of the substance whose molecular formula is C_2H_2 is *ethyne*. Its more common name is acetylene.

The structural formula for ethyne is _____.

$H-C\equiv C-H$

5-20   The general molecular formula for alkynes is C_nH_{2n-2}. The molecular formula for 3-octyne (oct = eight) will, therefore, be _____.

C_8H_{14}

5-21   The condensed structural formula for 3-octyne is _____.

$CH_3-CH_2-C\equiv C-CH_2-CH_2-CH_2-CH_3$

5-22   *Cyclic Hydrocarbons*
To form a ring from a carbon chain involves the removal of two hydrogen atoms from non-adjacent carbon atoms and the linking of the carbons thus exposed. *Cyclobutane* is

$$CH_2-CH_2$$
$$| \quad \quad |$$
$$CH_2-CH_2$$

Thus, cycloalkanes have the general formula C_nH_{2n} (as do alk*enes*).
The *molecular* formula of cyclopropane, therefore,

is _____, and its structural formula is _____.

C_3H_6 $CH_2 \overline{\quad\quad\quad} CH_2$
$\diagdown CH_2 \diagup$

5-23   The name for

$$CH_2-CH_2$$
$$| \quad\quad\quad CH_2$$
$$CH_2-CH_2 \diagup$$

would be _____

(remember the five sided building in Washington).

cyclopentane

5-24   Since cycloalk*anes* have the formula C_nH_{2n} , cycloalk*enes* (one double bond) have the formula C_nH_{2n-2}.
Cyclohexene, therefore, has the formula C_6H_____ .

Its structural formula is _____ .

---

10 (i.e., C_6H_{10})

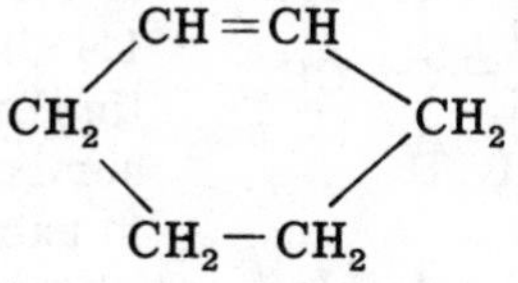

5-25   1,3-cyclohexadiene has the structural formula

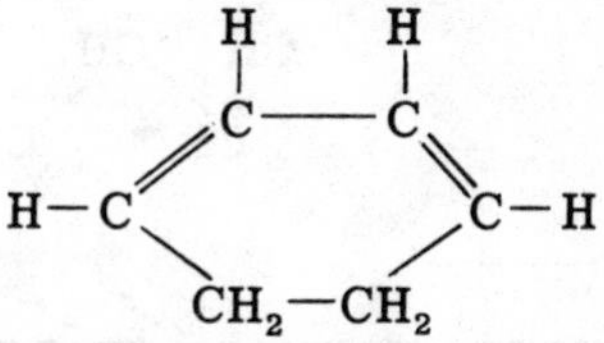

(Note that we count around the ring in such a way as
to keep the numbers as low as possible.)
    Benzene, C_6H_6, is often written as:

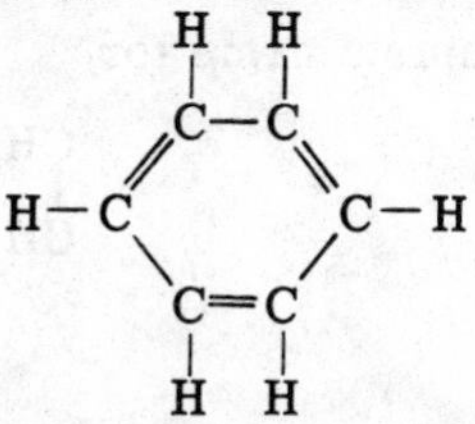

(though this does not give an adequate description of
its properties).

The name corresponding to this formula is

_____.

1,3,5-cyclohexatriene

5-26
$$H-C=C-CH_3$$
$$H-\underset{H}{\overset{|}{C}}-\underset{H}{\overset{|}{C}}-CH_3$$

is 1,4-dimethylcyclobutene or 1,4-dimethyl-1-cyclobutene. (In the absence of a number, the number 1 is assumed.) The ring is named first, and numbered to locate the double bond at the lowest numbered carbon atoms. The *ring* of

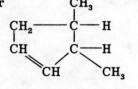

must be named

_____.

---

cyclopentene

5-27   The complete name for

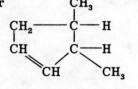

is

_____.

3,4-dimethylcyclopentene

5-28

is named _____.

5-ethyl-5,6-dimethyl-1,3-cycloheptadiene

The Systematic Naming of
Alkyl Halides

6-1 When a halogen atom is introduced into a hydrocarbon molecule, naming proceeds as if the halogen were simply a branch. The hydrocarbon is named as if the halogen were absent; the nature and position of the halogen is then indicated.

The *hydrocarbon* from which $CH_3CH_2CHCH_2CH_3$ is

derived is called _____.

Cl is under the CH.

---

pentane

6-2 $CH_3CH_2CHCH_2CH_3$ is called 3-chloropentane.
   |
   Cl

Analogously, $CH_3CH_2CH_2Cl$ must be named _____.

1-chloropropane

6-3 A Cl substituent is referred to as chloro-; Br substituents should then be named _____.

---

bromo-

6-4 CH_3CHFCH_3 is 2-fluoropropane. If I is called *iodo-*, $CH_3CH_2CH_2CHICH_2CH_3$ should be named

_____.

3-iodohexane

6-5   The name of the *hydrocarbon* from which

$$CH_3-CH-CH_2-CH-CH-CH_3$$

with CH_3 below the second carbon, Br below the fourth carbon, and CH_3 below the fifth carbon, is derived is

_____.

2,5-dimethylhexane

6-6 $(CH_3)_2CHCH_2CHBrCH(CH_3)_2$ is named 3-bromo-2, 5-dimethylhexane. (This is the same substance as shown in 6-5.) Substituents are listed in alphabetical order, regardless of their nature.

$CH_3CH_2CHClCH_2CHCH_3$ is _____.

with CH_3 below the fifth carbon.

4-chloro-2-methylhexane

6-7   If there are two or more atoms of the same halogen present, the prefixes di-, tri-, tetra-, etc. are used. 1,1,2-trichloropropane is

$$CH_3-CH-CH$$

with Cl below the second carbon, and Cl above and Cl below the third carbon.

CH_3-C-I is _____.

with I above and I below the central carbon.

1,1,1-triiodoethane

6-8 $ICH_2—CI_2—CH_2I$ is _____.

1,2,2,3-tetraiodopropane

6-9

$$CH_3—CH_2—\overset{\displaystyle Br}{\underset{\displaystyle Br}{\overset{|}{\underset{|}{CH}}}}—\overset{}{\underset{\displaystyle CH_3}{\overset{}{\underset{|}{CH}}}}—\overset{\displaystyle CH_3}{\underset{\displaystyle CH_3}{\overset{|}{\underset{|}{C}}}}——Cl \text{ is } \underline{\hspace{3cm}}.$$

(Remember to look for the longest carbon chain first.)

3,4-dibromo-2-chloro-2-methylhexane

6-10   If double bonds are present, the hydrocarbon is still named as if the halogen were absent. The halogen is named subsequently and the name is added to the hydrocarbon name. $F—CH=CH—CH_2—CH_3$ is 1-fluoro-1-butene.

      $CH_3—CH=CH—CH_2I$ is _____.

1-iodo-2-butene

6-11   1,3 butadiene $CH_2=CH—CH=CH_2$ on bromination yields

$$\underset{\displaystyle Br}{\overset{}{\underset{|}{CH_2}}}—CH=CH—\underset{\displaystyle Br}{\overset{}{\underset{|}{CH_2}}} \quad \text{and} \quad \underset{\displaystyle Br}{\overset{}{\underset{|}{CH_2}}}—\underset{\displaystyle Br}{\overset{}{\underset{|}{CH}}}—CH=CH_2$$

$$\qquad\qquad A \qquad\qquad\qquad\qquad\qquad B$$

Compound A would be named _____ and B

_____.

1,4-dibromo-2-butene   3,4-dibromo-1-butene

6-12   CH$_2$—CH=CH—CH=CH—CH$_2$ is named
             | |
             I   Cl

_____.

1-chloro-6-iodo-2,4-hexadiene

The Systematic Naming of Alcohols

7-1   If a hydrocarbon is characterized by the general formula R—H, an alcohol has the general formula R—OH. Alcohols may also be viewed as alkyl (R) substituted water molecules. The alcohol whose parent hydrocarbon is ethane, C_2H_6, has the formula _____.

C_2H_5OH

7-2   Halogen compounds were considered *substitution* products of hydrocarbons. Alcohols, on the other hand, are considered *functional* derivatives. The "functional group" OH is considered an integral part of the molecular structure for naming purposes:

C_2H_6 ethan*e* C_2H_5OH ethan*ol*
$CH_3CH_2CH_2CH_2CH_3$ pentan*e*
$CH_3CH_2CH_2CH(OH)CH_3$ 2-pentan*ol*
$CH_3CH_2CH_3$ propan*e*
$CH_3CH_2CH_2OH$ _____. (include position number)

1-propanol

7-3 $CH_3CH_2CH_2CH_3$
      butane $CH_3CH_2CH_2CH_2OH$
                                 _____

1-butanol

7-4 $CH_3CH_2CH_2CH_2CH_2CH_2CH_2CH_2OH$

1-octanol

7-5   Naming proceeds as for alkanes with one exception. Instead of looking for the longest carbon chain, we seek the longest carbon chain to which the OH group is directly attached.

$$\underset{\text{CH}_2\text{OH}}{\underset{1}{\overset{2}{\text{CH}_3}\overset{3}{\text{CH}_2}\overset{}{\text{CH}}\overset{4}{\text{CH}_2}\overset{5}{\text{CH}_2}\text{CH}_3}}$$

is 2-ethyl-1-pentanol (Note that it is *not* considered as a substituted hexane.)

The name of $CH_3CH_2CH_2\underset{CH_2CH_2OH}{CH}CH_2CH_2CH_2CH_3$ is _____.

3-propyl-1-heptanol

$$\overset{3\,4\,5\,6\,7}{\text{CCCCCCC}}$$
$$\underset{\underset{2\,1}{\text{CCOH}}}{\,}$$

7-6   Note that numbering on the parent alcohol chain is determined by the OH group. Its location is given the smallest possible number.

$$\underset{\text{OH}}{\overset{4}{\text{CH}_3}\overset{3}{\text{CH}_2}\overset{2}{\text{CH}}\overset{1}{\text{CH}_3}}$$

is 2-butanol not 3-butanol.

$CH_3CH_2CH_2CH(OH)CH_2CH_3$ is _____.

3-hexanol

7-7   Once the numbering system is determined by the location of the OH group, other branches, alkyl or halogen are named in the usual way.

$$CH_3-CH_2-\underset{\underset{CH_3}{|}}{\overset{\overset{CH_3}{|}}{C}}-OH \quad \text{is} \quad \underline{\hspace{4cm}}.$$

2-methyl-2-butanol

$$\overset{4}{C}-\overset{3}{C}-\underset{\underset{C}{|}}{\overset{\overset{\overset{1}{C}}{|}}{\overset{2}{C}}}-OH$$

7-8

$$\underset{\underset{CH_3}{|}\ \ CH_3CH}{CH_3CCH_2CH_2}\underset{}{\overset{\overset{CH_3}{|}\ \ \ \ \ \overset{CH_2CH_3}{|}}{C}}-CH_3$$

OH   is _____.

3-ethyl-3,6,6-trimethyl-2-heptanol

$$\overset{7}{C}-\overset{6}{\underset{\underset{C}{|}}{\overset{\overset{C}{|6}}{C}}}-\overset{5}{C}-\overset{4}{C}-\overset{3}{\overset{\overset{C-C}{|3}}{C}}-C$$

$$\overset{1}{C}-\overset{2}{\underset{OH}{C}}$$

7-9   *Polyhydroxy compounds*
    If possible, that carbon chain is chosen to which all OH groups are directly attached.

$$\underset{CH_2OH}{\overset{CH_2OH}{|}} \quad \text{is 1,2-ethanediol}$$

(common name is ethylene glycol)

$$CH_3CH_2 - \overset{\overset{\displaystyle CH_2OH}{|}}{\underset{\underset{\displaystyle CH_2OH}{|}}{C}} - OH \quad \text{is} \quad \underline{\hspace{4cm}}.$$

2-ethyl-1,2,3-propanetriol

7-10   2,2-dimethyl-1,3-propanediol has the structural

formula: _____.

$$CH_3 - \overset{\overset{\displaystyle CH_2OH}{|}}{\underset{\underset{\displaystyle CH_2OH}{|}}{C}} - CH_3$$

7-11   The insect repellant "6-12" (2-ethyl-1,3-hexane-diol) has the structural formula

_____.

$$\underset{\underset{\displaystyle OH \quad\quad\; OH}{|\quad\quad\;\; |}}{CH_2 CHCHCH_2CH_2CH_3}$$
$$\overset{\displaystyle CH_2CH_3}{|}$$

7-12   If it is impossible to find a continuous carbon chain to which all OH groups are attached, that chain is chosen to which most OH groups are attached. The remaining groups are named as *hydroxy* substituents.

$$CH_3-\underset{\underset{CH_2OH}{|}}{\overset{\overset{CH_2OH}{|}}{C}}-CH_2OH$$ is 2-(hydroxymethyl)-2-methyl-1,3-

propanediol. 4-(hydroxymethyl)-1,6-heptanediol has the formula

_____.

$$\underset{CCCCCCC}{\overset{\overset{CH_2OH}{|}}{\underset{OH}{}\quad |\quad \underset{OH}{}}}$$

OH   CH₂OH   OH

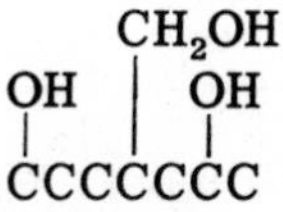

7-13   1,2-cyclopentanediol has the formula _____.

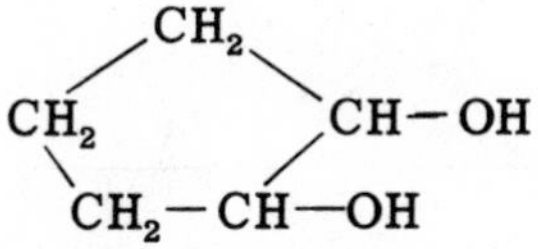

$$\begin{array}{c} CH_2 \\ CH_2 \qquad CH-OH \\ CH_2-CH-OH \end{array}$$

7-14

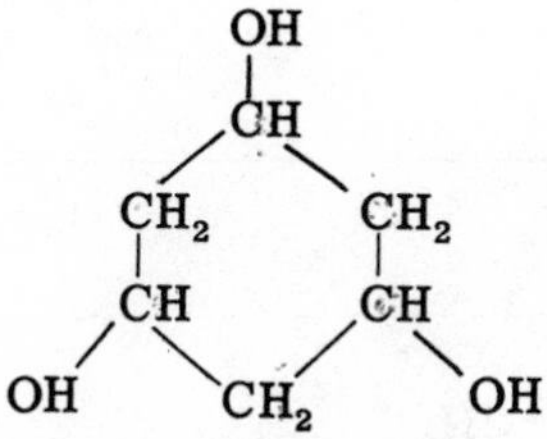

OH
CH
CH₂   CH₂
CH   CH
OH   CH₂   OH   is _____.

1,3,5-cyclohexanetriol

7-15   *Unsaturated alcohols*
$CH_3CH=CH-CH_2OH$ is 2-buten-1-ol
$CH_2=CH-CH_2OH$ is 2-propen-1-ol (note OH
(allyl alcohol) group takes precedence in
numbering)
3-buten-2-ol has the formula

_____ .

___

$CH_3-CH-CH=CH_2$
      $|$
    OH

___

7-16   If there is a choice of chains, the most unsaturated
is chosen as long as it still contains the OH group
(i.e., we line up again the maximum number of functional groups on the parent chain).
$CH_3CH=C-CH_2-OH$
         $|$
      $CH_2-CH_2-CH_3$ is _____ .

___

2-propyl-2-buten-1-ol

___

7-17   Remembering that compounds with triple bonds
are called alkynes, 2-propyn-1-ol has the formula

_____ .

___

$HC\equiv C-CH_2OH$

___

7-18 $CH_3CH_2C-CH_2-CH-CH_3$
         $||$ $|$
       CH_2 CH_2OH is _____ .

___

4-ethyl-2-methyl-4-penten-1-ol

___

7-19   The formula of 2,2,5-trimethyl-3-hexene-1,5-diol
is

———————————— .

$$\underset{\substack{| \\ OH}}{CH_2} - \underset{\substack{| \\ CH_3}}{\overset{\overset{CH_3}{|}}{C}} - CH = CH - \underset{\substack{| \\ OH}}{\overset{\overset{CH_3}{|}}{C}} - CH_3$$

The Systematic Naming of Aldehydes and Ketones

8-1   Aldehydes and ketones may be considered as derived from hydrocarbons by replacing one CH_2 group by a carbonyl group, $C=O$. If the $C=O$ group is at the end of a chain or branch we have an aldehyde, if flanked by a carbon on each side, a ketone. Which of the following (A, B, C, etc.) are aldehydes?

(A) $CH_3\overset{\|}{\underset{O}{C}}H$,   (B) $CH_3CH_2\overset{\|}{\underset{O}{C}}CH_3$,

(C) $(CH_3)_2CHCOCH(CH_3)_2$, (D) HCHO, (E) CH_3COCH_3,
(F) $HCOC_2H_5$

A, D, F

8-2

$$\underset{\text{Methane}}{H-\overset{\displaystyle H}{\underset{\displaystyle H}{C}}-H} \qquad \underset{\text{Methanal}}{\overset{\displaystyle H-C-H}{\underset{\displaystyle O}{\|}}} \qquad \underset{\text{Ethane}}{H-\overset{\displaystyle H}{\underset{\displaystyle H}{C}}-\overset{\displaystyle H}{\underset{\displaystyle H}{C}}-H} \qquad H-\overset{\displaystyle H}{\underset{\displaystyle H}{C}}-\overset{\|}{\underset{O}{C}}-H$$

Methane   Methanal   Ethane   _____

Ethanal

8-3   propane $CH_3CH_2CH_3$ propanal _____

$CH_3CH_2\overset{\|}{\underset{O}{C}}H$ Note: No position number is needed, because CHO group is always considered as position 1.

8-4

$$H-\overset{\overset{\displaystyle H}{|}}{\underset{\underset{\displaystyle H}{|}}{C}}-\overset{\overset{\displaystyle H}{|}}{\underset{\underset{\displaystyle H}{|}}{C}}-\overset{\overset{\displaystyle H}{|}}{\underset{\underset{\displaystyle H}{|}}{C}}-\overset{\overset{\displaystyle H}{|}}{\underset{\underset{\displaystyle H}{|}}{C}}-H = CH_3CH_2CH_2CH_3 = CH_3(CH_2)_2CH_3$$

= butane

$$H-\overset{\overset{\displaystyle H}{|}}{\underset{\underset{\displaystyle H}{|}}{C}}-\overset{\overset{\displaystyle H}{|}}{\underset{\underset{\displaystyle H}{|}}{C}}-\overset{\overset{\displaystyle H}{|}}{\underset{\underset{\displaystyle H}{|}}{C}}-\overset{\overset{\displaystyle H}{|}}{C}=O = CH_3CH_2CH_2CHO = CH_3(CH_2)_2CHO$$

= _____

butanal

8-5 $CH_3CH_2CH_2CH_2CH_2CH_2CH_2CHO$ = _____

octanal

8-6

$CH_3CH_2\overset{}{\underset{\underset{\displaystyle CH_3}{|}}{C}}HCHO$ 2-methylbutanal

$CH_3CH_2\overset{}{\underset{\underset{\displaystyle C_2H_5}{|}}{C}}HCH_2CHO$

_____

3-ethylpentanal

8-7 2-chloropropanal _____

$CH_3-\overset{}{\underset{\underset{\displaystyle Cl}{|}}{C}}H-CHO$

8-8   2,2,2-trichloroethanal
       (common name chloral)

_____

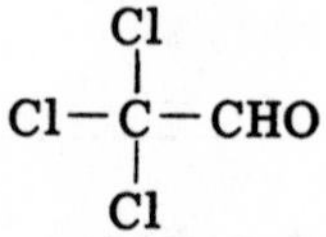

$$Cl-\underset{\underset{Cl}{|}}{\overset{\overset{Cl}{|}}{C}}-CHO$$

8-9 $CH_3-\underset{\underset{O}{\|}}{C}-CH_3$ propanone—a ketone

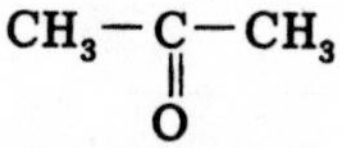

       (common name acetone)

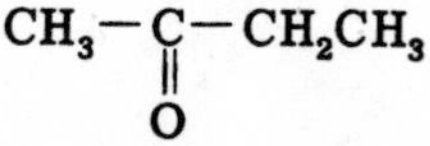

       $CH_3-\underset{\underset{O}{\|}}{C}-CH_2CH_3$

       _____

       butanone (Note: No location number is needed as
                 the name is unambiguous.)

8-10 $CH_3-\underset{\underset{O}{\|}}{C}-CH_2-CH_2-CH_3$ $CH_3-CH_2-\underset{\underset{O}{\|}}{C}-CH_2-CH_3$

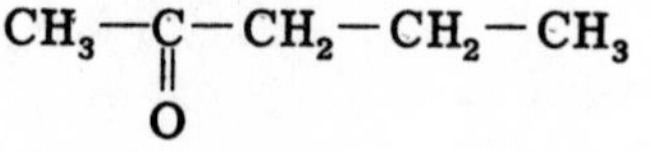

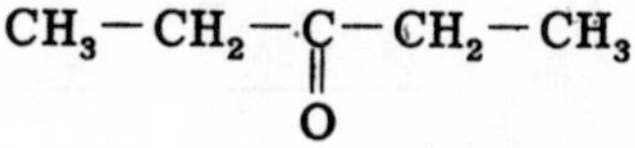

       2-pentanone

                                                            _____

       3-pentanone

8-11 CH_3
            $|$
       $CH_3\underset{\underset{O}{\|}}{C}CHCH_2CH_3$

                                          _____

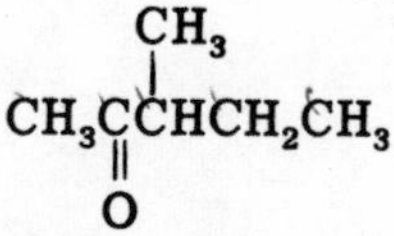

       3-methyl-2-pentanone

8-12

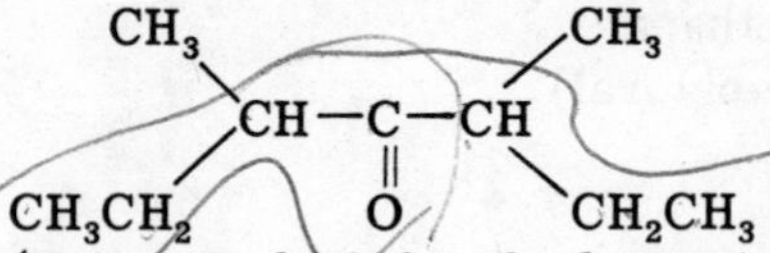

$$CH_3-CH-C-CH-CH_3$$

with CH_3CH_2 and O and CH_2CH_3 is _____.

(Note: We look for the longest chain containing the CO group.)

3,5-dimethyl-4-heptanone

8-13 $CH_3-\underset{\underset{O}{\|}}{C}-CH_2-\underset{\underset{O}{\|}}{C}-CH_3$ is 2,4-pentanedione.

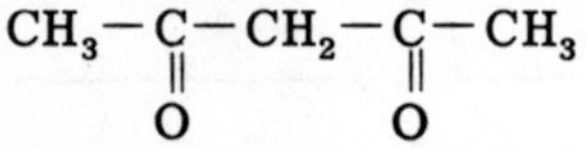

2,3-hexanedione is _____.

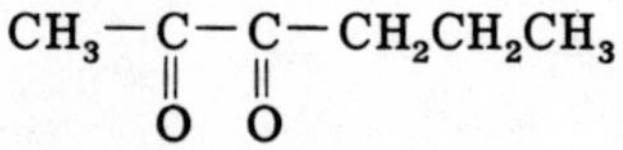

$CH_3-\underset{\underset{O}{\|}}{C}-\underset{\underset{O}{\|}}{C}-CH_2CH_2CH_3$

8-14 1,4-cyclohexanedione is

_____

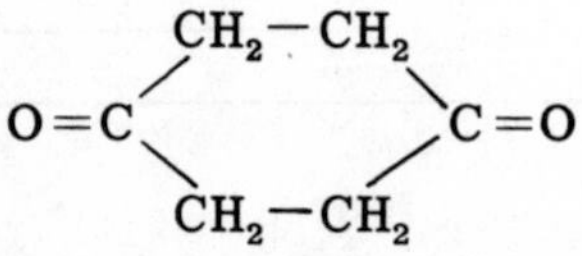

$O=C\underset{CH_2-CH_2}{\overset{CH_2-CH_2}{\diagup\diagdown}}C=O$

8-15 $CH_3-CH=CH-CHO$ is 2-butenal

3-penten-2-one is _____ .

$$CH_3-\underset{\underset{O}{\|}}{C}-CH=CH-CH_3$$

8-16   C=O takes precedence over OH in determining the parent name thus: $CH_3CH_2\underset{\underset{O}{\|}}{C}CH_2CH_2OH$ is 1-hydroxy-3-pentanone, and 3-hydroxypentanal is _____ .

$$CH_3CH_2\underset{\underset{OH}{|}}{C}HCH_2CHO$$

8-17   5-chloro-3-heptenal is _____ .

$$CH_3CH_2-\underset{\underset{Cl}{|}}{C}H-CH=CH-CH_2-CHO$$

8-18 $CH_3-\underset{\underset{O}{\|}}{C}-CH_2-\underset{\underset{Cl}{|}}{C}H-CH_3$ is _____ .

4-chloro-2-pentanone

8-19   2,5-cyclohexadien-1-one has the structural formula

_____

$$CH_2 \underset{CH=CH}{\overset{CH=CH}{<}} C=O$$

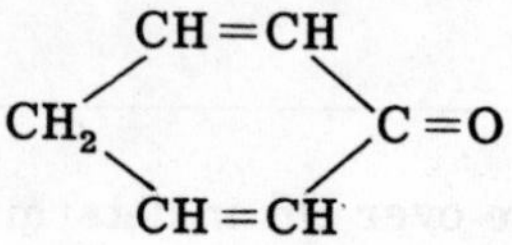

8-20   2,5-heptadien-4-one has the structural formula

_____

$$CH_3-CH=CH-\underset{\underset{O}{\|}}{C}-CH=CH-CH_3$$

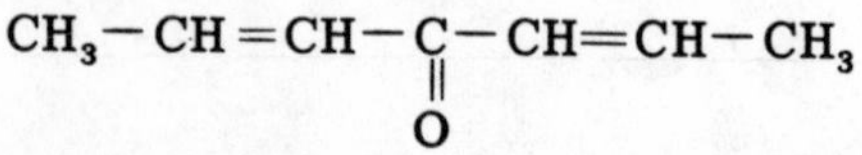

Some Comments on the Common Names of Carboxylic Acids, Aldehydes and Ketones

PART A—CARBOXYLIC ACIDS

These contain the functional group $-\overset{\displaystyle O}{\underset{\displaystyle \|}{C}}-OH$, often abbreviated $-CO_2H$ (and sometimes $-COOH$).

The traditional names of carboxylic acids follow few rules except that they end in *-ic acid*. They had best be quickly memorized. A few aids to memorization are given here.

9-1   Formic acid, the smallest carboxylic acid, HCO_2H, is found in red ants and can be obtained from them by distillation or by being bitten by them. The French for ant is *la fourmi* and the Latin is *formica*. Hence the name. Formica table tops are made from a phenol-formaldehyde resin; formaldehyde, $HCHO$, is a reduction product of formic acid.

The acid, CH_3CO_2H, is the major constituent of vinegar. Since the Latin for vinegar is *acetum*, the name of the acid is _____.

———————————————————————————

acetic acid.

———————————————————————————

9-2   Propionic acid, $CH_3CH_2CO_2H$, is related to propane because they contain the same number of _____

_____ .

carbon atoms

9-3   The next, butyric acid, $CH_3CH_2CH_2CO_2H$, is the acid that supplies the unpleasant smell in rancid butter, and derives its name from that fact. The molecule contains four carbon atoms, as does the alkane

_____ .

butane (C_4H_{10})

9-4   *Valerus* is Latin for strong. The substance whose formula is $CH_3CH_2CH_2CH_2CO_2H$ has a strong smell. Hence the 5-carbon acid is likely to be known as

_____ .

valeric acid.

9-5   The ancient Romans called a goat *caper*. The six-, eight-, and ten-carbon acids smell of goats and are accordingly named caproic, caprylic, and capric acid. Since few people can remember which is which, the IUPAC names, hexanoic, octanoic, and decanoic acids are preferred. Two acids were left out—the seven-carbon acid, $C_6H_{13}CO_2H$, known as *n*-heptylic acid, and the nine-carbon acid, of formula _____ , and

named _____ (nona = 9).

$C_8H_{17}CO_2H$ *n*-nonylic acid

9-6   *To review:* La Fontaine wrote a fable of the ant and the grasshopper, entitled, *"La Cigale et la Fourmi."* HCO_2H is named _____.

---

formic acid

9-7   The two-carbon acid constituent of vinegar has the name _____ and the formula is _____.

---

acetic acid CH_3CO_2H

9-8   The four-carbon acid, $C_3H_7CO_2H$, is named _____.

---

butyric acid

9-9   Valeric acid has the formula _____.

$C_4H_9CO_2H$

PART B—ALDEHYDES

9-10   If CH_3CO_2H is acetic acid, then CH_3CHO is acetaldehyde. Propionaldehyde by analogy, must have the formula _____.

CH_3CH_2CHO

9-11 And HCHO will be named _____.

---

formaldehyde

9-12 There are two butyraldehydes (as there are two butyric acids) one known as butyraldehyde or *n*-butyraldehyde, the other as isobutyraldehyde. Their

formulas are _____ and _____.

$CH_3CH_2CH_2CHO$ *n*-butyraldehyde

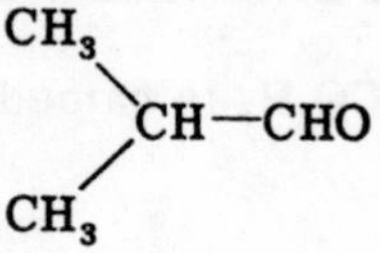

$$\begin{matrix} CH_3 \\ \quad\diagdown \\ \quad\quad CH-CHO \quad \text{isobutyraldehyde} \\ \quad\diagup \\ CH_3 \end{matrix}$$

PART C—KETONES $R-\underset{\underset{O}{\|}}{C}-R'$

9-13 The common names of ketones are very simple. The two groups R and R' are named by their common names in alphabetical order followed by the word ketone. The name is written as three separate words. The name for $CH_3-\underset{\underset{O}{\|}}{C}-C_2H_5$, therefore, must be

_____.

---

ethyl methyl ketone

9-14 isopropyl *n*-propyl ketone has the skeleton formula

_____.

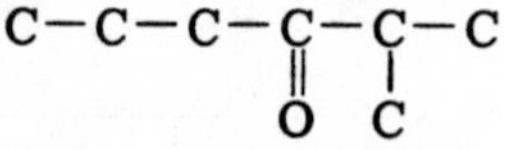

$$C-C-C-\underset{\underset{C}{|}}{\overset{\overset{O}{\|}}{C}}-C-C$$

9-15 Dimethyl ketone (which has an even commoner name, acetone) has the formula

_____.

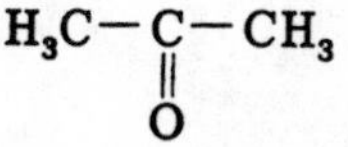

$H_3C-\underset{\overset{\|}{O}}{C}-CH_3$ or $C-\underset{\overset{\|}{O}}{C}-C$ or $(CH_3)_2CO$

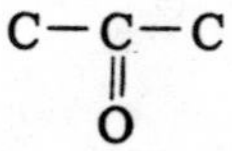

9-16 $C-C-C-C-\underset{\overset{\|}{O}}{C}-C$ is _____.

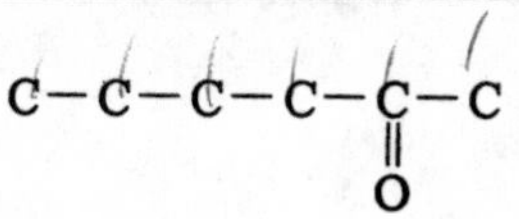

n-butyl methyl ketone

9-17 $C-\underset{\underset{C}{|}}{C}-\underset{\overset{\|}{O}}{C}-C-C$ is _____.

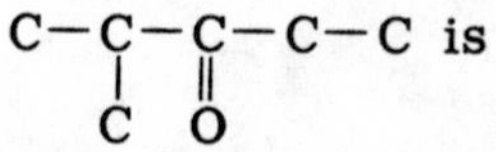

ethyl isopropyl ketone

9-18 $C-C-C-\underset{\overset{\|}{O}}{C}-C-C-C$ is _____.

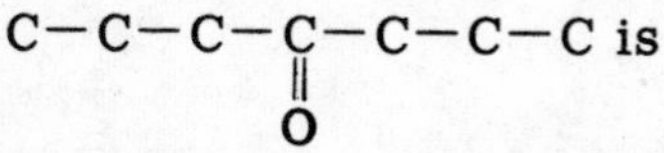

di-*n*-propyl ketone

9-19   ethyl isopentyl ketone has the skeleton formula

_____.

$$C-C-\underset{\underset{O}{\|}}{C}-C-C-C\begin{smallmatrix}C\\ \\C\end{smallmatrix}$$

9-20   And diisopropyl ketone is _____.

$$C-\underset{\underset{C}{|}}{C}-\underset{\underset{O}{\|}}{C}-\underset{\underset{C}{|}}{C}-C$$

The Systematic Naming of Carboxylic Acids

10-1

$$H-\overset{\overset{\displaystyle H}{|}}{\underset{\underset{\displaystyle H}{|}}{C}}-\overset{\overset{\displaystyle H}{|}}{\underset{\underset{\displaystyle H}{|}}{C}}-H \qquad H-\overset{\overset{\displaystyle H}{|}}{\underset{\underset{\displaystyle H}{|}}{C}}-\overset{\overset{\displaystyle }{|}}{\underset{\underset{\displaystyle O}{||}}{C}}-OH \quad \text{or} \quad CH_3CO_2H$$

    ethane   ethanoic acid

$CH_3CH_2CH_3$ _____

   propane   propanoic acid

$$H-\overset{\overset{\displaystyle H}{|}}{\underset{\underset{\displaystyle H}{|}}{C}}-\overset{\overset{\displaystyle H}{|}}{\underset{\underset{\displaystyle H}{|}}{C}}-\overset{\overset{\displaystyle }{|}}{\underset{\underset{\displaystyle O}{||}}{C}}-OH \quad (\text{or } CH_3CH_2CO_2H \text{ or } C_2H_5CO_2H)$$

10-2   Pentanoic acid is _____.

$CH_3CH_2CH_2CH_2CO_2H$ $(= C_4H_9CO_2H)$ (*Not* $C_5H_{12}CO_2H$; the carbon in the CO_2H group is considered *part* of the original pentane.)

10-3 3-hydroxybutanoic acid is _____.

_____. Note: The carbon of the CO_2H group can only be at the end of a carbon chain, never in the middle. Thus it is considered as carbon atom number 1, taking precedence over ketone or alcohol functions.

$CH_3CHCH_2CO_2H$
　　|
　　OH

10-4 2-but*e*noic acid is _____.

$CH_3CH=CHCO_2H$

10-5 "Acrylic acid" has the formula $CH_2=CHCO_2H$. Its Systematic name is _____.

---

prop*e*noic acid

10-6 $CH_3CHCH_2CO_2H$ is 3-methylpentanoic acid
　　　|
　　CH_3CH_2

$(CH_3CH_2)_3CCO_2H$ is _____.

2,2-diethylbutanoic acid
　　　　　　　　　　C
　　　　　　　　　　C
　　　　　　　CCCCO₂H
　　　　　　　　　　C
　　　　　　　　　　C

10-7 $CH_3-CH_2-\underset{\underset{CO_2H}{|}}{CH}-CH_2-CH_3$ is _____ .

2-ethylbutanoic acid

10-8   3,5-hexadienoic acid is _____ .

$CH_2=CH-CH=CHCH_2CO_2H$

10-9 $\underset{CH_2CO_2H}{\overset{CH_2CO_2H}{|}}$ is butanedioic acid.

$CH_2(CO_2H)_2$ is _____ .

propanedioic acid (also known as malonic acid)

10-10   3-hydroxypentanoic acid is _____ .

$\underset{\underset{OH}{|}}{CH_3CH_2CH}CH_2CO_2H$

Esters — Common and Systematic Names

11-1   An ester can usually be decomposed by water (catalyzed in most cases by acids or bases) to form an alcohol and a carboxylic acid. The names of esters reflect this relationship. Thus an ester which on hydrolysis forms 1-propanol and ethanoic acid is named propyl ethanoate. Ethyl butanoate when allowed to react with water would produce

_____ and _____ acid.

---

ethanol   butanoic acid

11-2   You will notice that the alcohol part of the ester is always named first, and the acid part (containing the $C = O$ group) second. Unfortunately the *formulas* are usually written with the acid part first. Thus the formula of methyl propanoate is usually written $C_2H_5C-OCH_3$, or $C_2H_5CO_2CH_3$, or even $C_2H_5COOCH_3$.
        $\overset{\|}{O}$

Our problem then is to unscramble the formula to be sure which part belongs to the acid and which to the alcohol. Let us look at the equation for an ester synthesis:

$$R-\underset{\underset{O}{\|}}{C}-O-H + R'-O-H \xrightarrow[\text{catalysis}]{\text{acid}} R-\underset{\underset{O}{\|}}{C}-O-R' + H_2O$$

83

Note that there is a difference in the bonding of the
R and R′ groups. The R group throughout the reac-
tion is attached to a(n) _____ atom, whereas
R′ in alcohol and ester is attached to a(n)
_____ atom.

---

carbon   oxygen

11-3   In the ester formula, then, the alkyl group attached to carbon belongs to the acid portion, whereas the alkyl group attached to oxygen came from
the alcohol. If we unscramble the formula
$CH_3CO_2C_2H_5$ to show all bonds, we obtain the structural formula

_____.

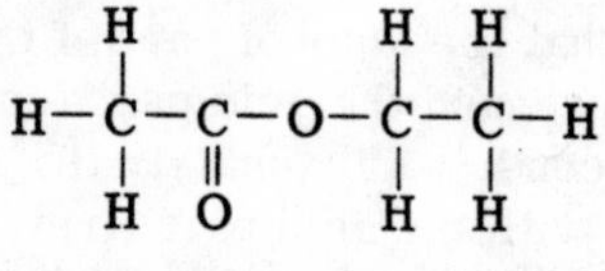

11-4   We may now write the formulas of the acid and
the alcohol from which this ester $CH_3CO_2C_2H_5$ was
derived. The formula of the acid was _____
and of the alcohol _____.

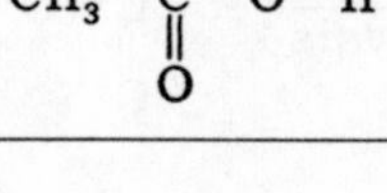

11-5   The Systematic (IUPAC) names of these two substances are _____ and _____.

---

ethanoic acid   ethanol

11-6   The name of the ester obtained from these two
substances is therefore _____.

---

ethyl ethanoate

11-7   The name of $CH_3CH_2CH_2CO_2CH_3$ is _____.

---

methyl butanoate

11-8   If the constituent alcohol is 1-pentanol and the
acid is 2-methylhexanoic acid, the ester will be
named pentyl 2-methylhexanoate. Its formula must
be

_____

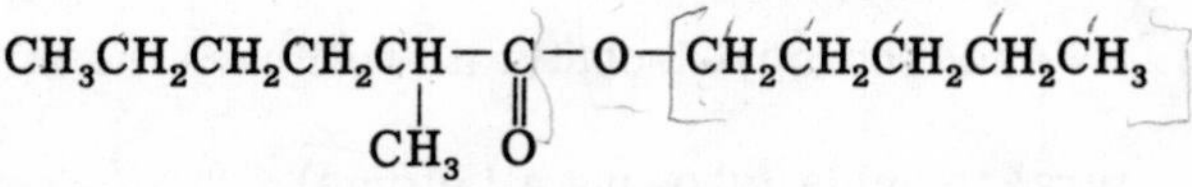

$$CH_3CH_2CH_2CH_2CH-\overset{\displaystyle}{\underset{\displaystyle CH_3}{C}}-O-\left[CH_2CH_2CH_2CH_2CH_3\right]$$
$$\overset{\displaystyle O}{}$$

11-9 $CH_3CH_2-\underset{\displaystyle O}{\overset{\displaystyle}{C}}-O-CH_2CH_2CH_2CH_3$ is named

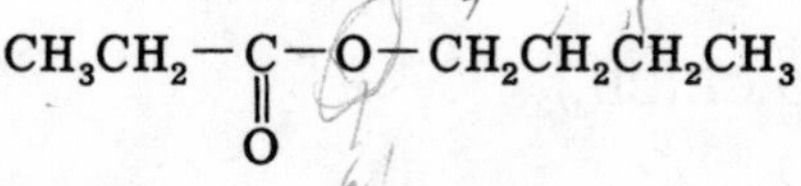

_____

---

butyl propanoate

11-10 $CH_3(CH_2)_4CHCO_2CH_2CH_2CH_3$ is named
$\qquad\qquad\qquad$ CH_2CH_3

_____ .

propyl 2-ethylheptanoate

11-11   We will briefly indicate here the modifications to be made if common rather than IUPAC names are desired. The major principle remains the same. The alkyl group of the alcohol is named first, followed by part of the acid name changed to end in -ate. The alcohol and acid names used are of course their common names. Thus the ester from ethyl alcohol and acetic acid will be named _____ .

ethyl acetate

11-12   The ester from propionic acid and *tert*-butyl alcohol will have the common name _____ .

tert-butyl propionate

11-13   And isopropyl isobutyrate will have the structural formula (showing all atoms) _____ .

$CH_3-CH-CO_2CH(CH_3)_2$
$\qquad\ \ CH_3$

11-14   The common and IUPAC names for an ester
whose skeletal structure is $C-C-C-CO_2C-C-C$

are _____ and _____ .

---

common name: *n*-propyl *n*-butyrate
IUPAC name: propyl butanoate

An Introduction to Aromatic Nomenclature

Compared with the problems of aliphatic nomenclature discussed so far, the nomenclature of substituted benzene derivatives is extremely simple. Since the benzene ring is generally considered to be a plane regular hexagon, a monosubstituted benzene derivative does not need to have its position specified:

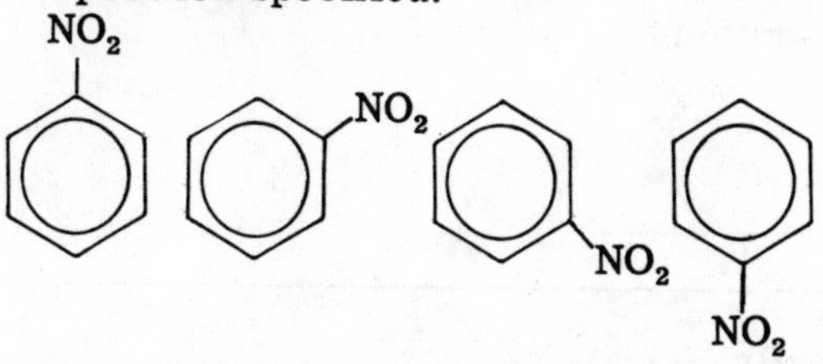

. . ., etc. = nitro-benzene

12-1   With two nitro group substituents, there are several possibilities. First, they may be adjacent to each other:

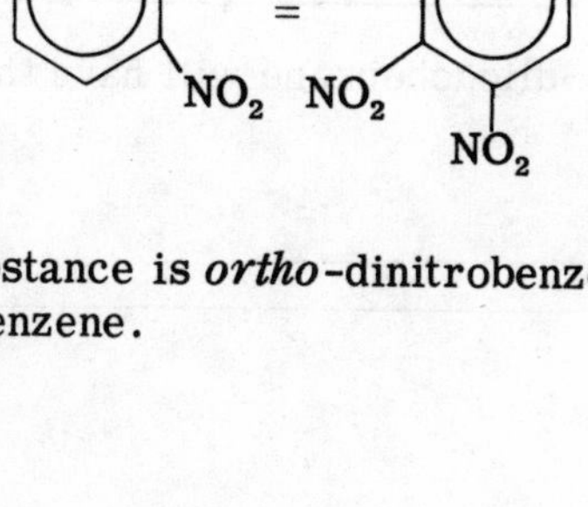

This substance is *ortho*-dinitrobenzene, written *o*-dinitrobenzene.

The second possibility has one carbon between the two nitrosubstituted carbon atoms thus

_____ .

12-2   The name for this substance is *meta*-dinitrobenzene, written *m*-dinitrobenzene. *m*-Dibromobenzene can then be written

_____ .

(or equivalent)

12-3   and *m*-diiodobenzene will have the formula

_____ .

12-4 and *o*-dibromobenzene will have the formula

_____ .

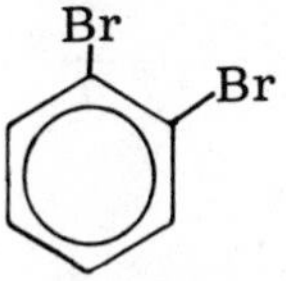

12-5 c) In addition to ortho- and meta-disubstitution there $\frac{\text{is one}}{\text{are two}}$ more possible arrangement(s).

is one more

12-6 This third arrangement (for dinitrobenzene) is

_____

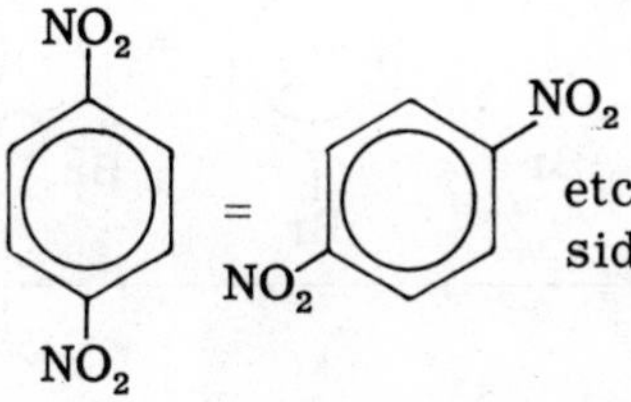

etc. with groups on *opposite* sides of the ring.

12-7   and is known as *para*-dinitrobenzene or *p*-dinitro-
benzene. Suppose that *p*-dibromobenzene, whose
formula is

_____

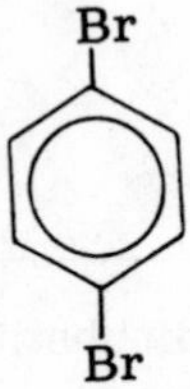

12-8   is further brominated to form a tribromobenzene.
There are four unsubstituted carbon atoms in di-
bromobenzenes. How many *different tri*bromobenzenes can be prepared from *p*-dibromobenzene?

_____ .

Only one because

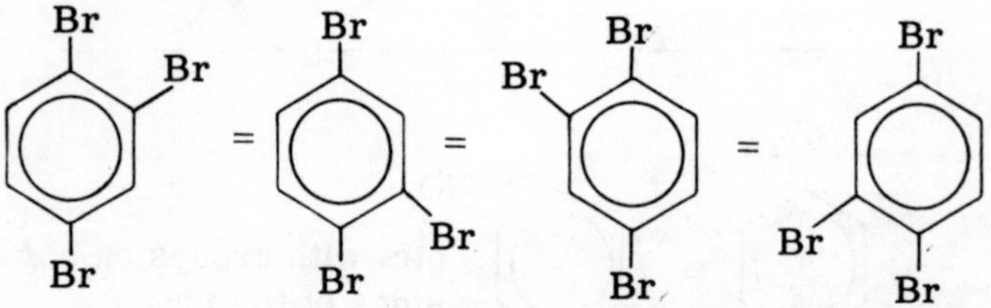

12-9   With *o*-dibromobenzene, whose formula is

_____

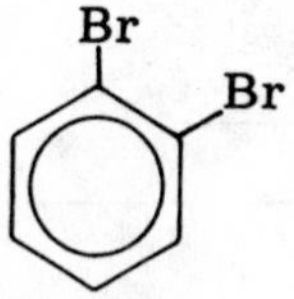

12-10   the number of different tribromobenzenes that can
be made by introducing one more bromine into the
ring is _____.

two

12-11   These have the structures _____ and

_____.

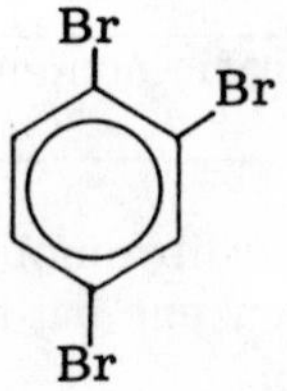

and

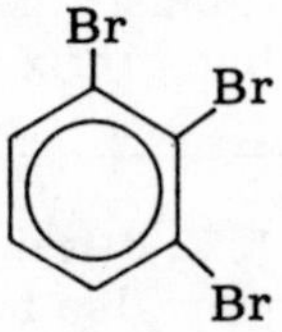

I   II

Be sure to recognize that any other formulas you
have drawn are equivalent to I or II or are in error.

12-12   Repeating with *m*-dibromobenzene we find we can

form _____ different tribromobenzenes
　　　　　　(number)

which have formulas

_____

three   Br   Br   Br

I   Br   Br   Br   Br
                              II   III

12-13   We find that I, II, and III represent the only tribromobenzenes that can be made. They are named
according to the following rule: One of the bromines
is designated as attached to carbon atom number 1.
Numbering is now carried on around the ring in that
direction which allows the other two bromines to
have the smallest position numbers possible. That
bromine is chosen as number one that gives the
smallest numbers for all three bromines. Formula
II must, therefore, be named _____
(the prefix *tri*- is used as in aliphatic nomenclature).

_____

1, 2, 3-tribromobenzene   Note: *o*-dibromobenzene
may also be called 1,2-dibromobenzene but this is
less common.

_____

12-14   The name for

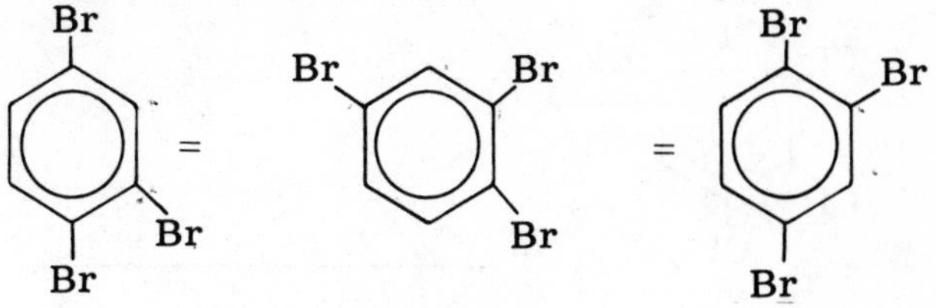

is _____ .

1, 2, 4-tribromobenzene

12-15   The symmetrical tribromobenzene, III (some-times known as *sym*-tribromobenzene), will be named _____ .

1,3,5-tribromobenzene

12-16   1,2,4,5-tetramethylbenzene has the formula

_____ .

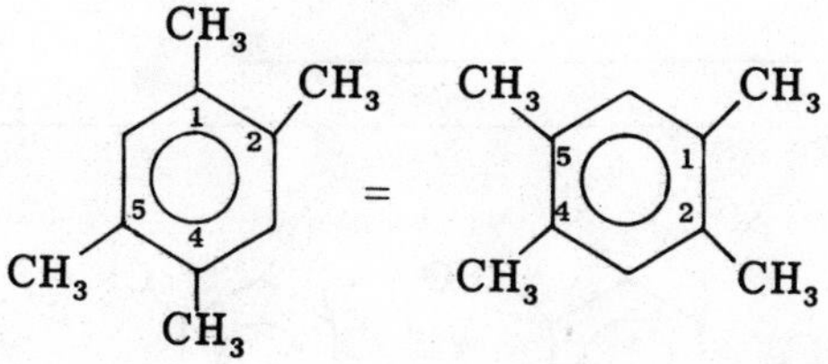

12-17

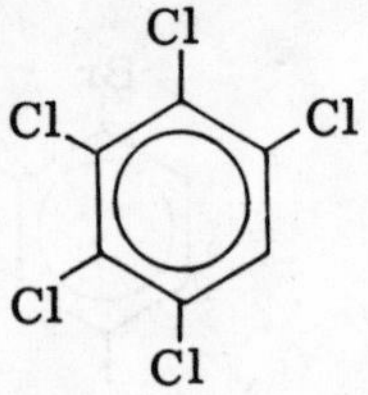

is _____ .

1,2,3,4,5-pentachlorobenzene

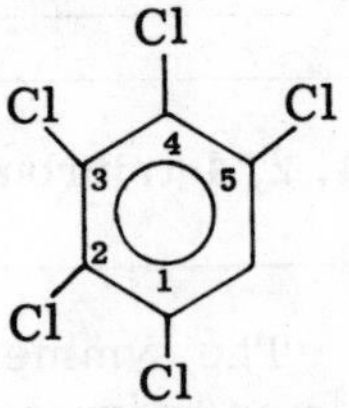

12-18   When two or more different groups are substituted in benzene, an order of precedence must be
established. Some monosubstituted benzene derivatives have special names such as aniline
($C_6H_5NH_2$), phenol (C_6H_5OH), benzaldehyde (C_6H_5CHO),
and benzoic acid ($C_6H_5CO_2H$). The functional group
in these substances is considered as attached to
carbon atom number one. 3-chlorobenzaldehyde,
therefore, has the formula

_____ .

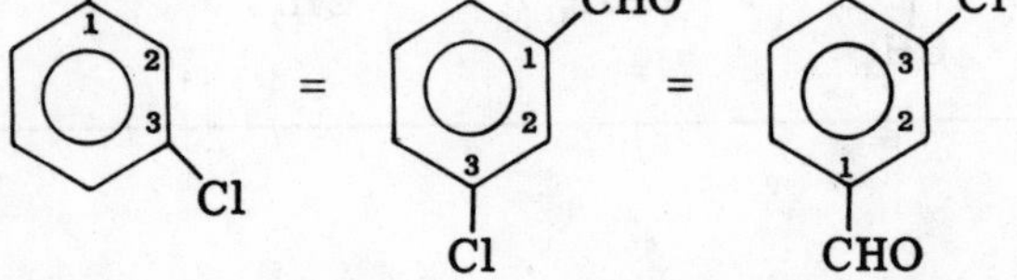

12-19   3,5-dinitrobenzoic acid is represented as

_____

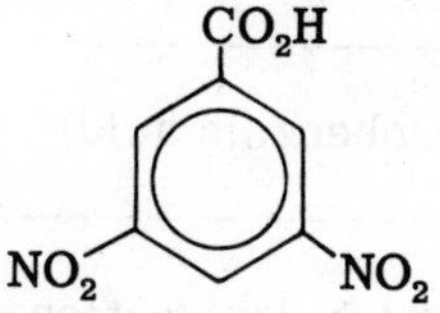

12-20   and 2,4-dinitroaniline as

_____ .

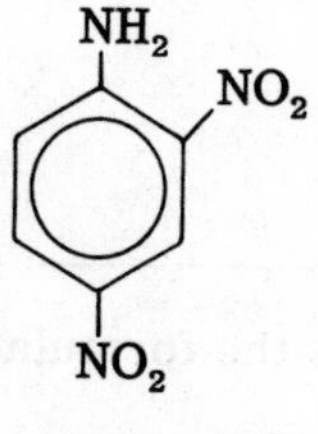

12-21

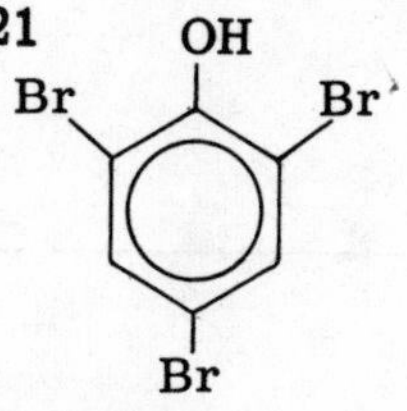

is _____ .

2,4,6-tribromophenol

12-22      is _____.

4-chlorobenzoic acid (or *p*-chlorobenzoic acid)

12-23   On the basis of the *o*-, *m*-, and *p*-designation dis-
         cussed earlier, *m*-chlorobenzoic acid would have
         the formula

         _____.

12-24 1-chloro-2,4-dinitrobenzene has the formula

         _____.

12-25   and   NO$_2$
                    Br

                    NO$_2$ would be named _____ .

2-bromo-1,3-dinitrobenzene

12-26 In heterocycles such as pyridine,

the heteroatom (in this case, N), is taken to occupy
the number one position. 2,4-dimethylpyridine thus
has the formula

_____ .

CH$_3$

CH$_3$ N

12-27 And
       Br   Br

       Br   N   Br   is named _____ .

2,3,5,6-tetrabromopyridine

The Calculation of Formal Charges

The next two chapters represent a detailed analysis of the information derivable from an electronic formula, the procedure for constructing such formulas and two alternate ways of depicting them. In the course of construction of more complex formulas (CO_2, $COCl_2$, HNO_3), the problem of multiple formulas for the same compound is encountered, serving as an introduction to those questions dealt with by the theory of resonance. This topic will be dealt with in Chapter 15. The present chapter also introduces the concept of "kernel charge," and emphasizes the fact that the symbols for the elements in extended electronic or line formulas represent not the nuclei but the kernels of the atoms, that is, the nuclei plus the completed inner electron shells of isolated atoms.

In predicting the properties of a substance with a given structural formula, it is of great importance to know the location of particularly large concentrations of positive or negative charge as these will strongly influence other mole- cules in their immediate environment. This in turn largely determines such properties as boiling point, melting point, density and the ease of breaking of some chemical bonds.

The term "formal" charge is used because its calcula- tion gives only a first approximation to the actual charge distribution in a molecule. We shall develop a formalism according to which trimethylamine oxide may be written

$$
\begin{array}{ccccc}
 & & H & & \\
 & & | & & \\
H & H\!-\!C\!-\!H & & \\
| & | & & \\
H\!-\!C\!-\!\!\!-\!\!\!-\!N^{\oplus}\!\!-\!\!\!-\!\!\!-\!\overline{\underline{O}}|\;^{\ominus} \\
| & | & & \\
H & H\!-\!C\!-\!H & & \\
 & & | & & \\
 & & H & & \\
\end{array}
$$

The formula shows a formal charge of +1 for nitrogen, −1 for oxygen and zero for all hydrogen and carbon atoms.

To calculate formal charges, we make one major over-simplification; namely, that shared electrons are shared equally. This is obvious in H:H, it is most unlikely in H:F, knowing its acid character. (For a better approximation we would need to incorporate some measure of the relative electron attracting power of different atoms, or more technically, of their relative electronegativities. This factor is ignored in formal charge calculations.)

Our procedure, then, is to split all shared electron groups into two equal parts and calculate the formal charge —that is, the net charge around each nucleus.

13-1   Ammonia H : N̈ : H would be split H ·|· N̈ ·|· H

         H   H
              H

Similarly, water H : Ö : would be split _____.

         H
         ·
H ·|· Ö :
         ··

13-2   We ask what positive and negative charges exist around each nucleus. The net charge (that is, the algebraic sum of all positive and negative charges) represents the formal charge on the atom in the given molecule.

The hydrogen nucleus has one proton with a charge of +1. If associated with one electron (charge −1) its formal charge is (+1 −1) or 0. A hydrogen nucleus associated with two electrons (H:) would have a formal charge of _____.

−1 (i.e., +1 −2)

13-3   Carbon has an atomic number of 6. There are, therefore, 6 protons in the nucleus, and its nuclear charge is +6. Further, in all its compounds, carbon carries an inner shell of two electrons or a charge of −2. We speak of the "kernel" of an atom as the nucleus plus the inner shells of electrons. The kernel of carbon has an overall charge of

_____.

+4 (i.e., +6 −2)

13-4   If the carbon kernel of charge +4, containing 6 protons and 2 electrons (neutrons are of no significance in charge calculations), is associated with four further electrons, the formal charge would be

_____.

---

0 (i.e., + 6 −2 −4)

13-5   If the carbon kernel had 3 electrons around it, the formal charge would be _____.

+1 (i.e., +6 (in nucleus) $\Big\}$ the kernel
$\quad\quad\quad \underline{-2}$ (in inner shell)
$\quad\quad\quad +4$
$\quad\quad\quad \underline{-3}$ (electrons outside kernel)
$\quad\quad\quad +1$

13-6   The elements belonging to the first period of eight in the periodic table carry a completed inner shell of two electrons and have atomic number Li = 3, Be = 4, B = 5, C = 6, N = 7, O = 8, F = 9. The kernel charges of the atoms from Li to F will be _____.

+1, +2, +3, +4, +5, +6, +7

13-7   The ammonia molecule (Frame 13-1), was written
H·|· N̈ ·|· H. Each hydrogen has one electron in its
          ·
          H

immediate environment. Given that formal charge
= number of protons minus number of electrons, the
formal charge on H is _____.

0

13-8   H·|· N̈ ·|·H   The atomic number of nitrogen is 7,
          ·   the number of protons in the nucleus
          H   is, therefore, _____; the number of
electrons in its inner shell is _____; and the
kernel charge (including sign) is _____.

---

7   2   +5

13-9   H·|· N̈ ·|· H   In the formula of ammonia, the symbol
          ·   N stands for the *kernel* of the nitrogen
          H   atom, not merely for the nitrogen nu-
cleus. Chemical formulas almost invariably show
only the outer electrons, that is, electrons in the
outer shell. The kernel charge of nitrogen is
_____ in the formula above, and the number of
electrons associated with the kernel (but outside it)
is _____.

+5   5

13-10 H ·|· Ṅ ·|· H   The formal charge on the nitrogen atom
      Ḣ   is equal to the number of protons in the
                   nucleus minus the number of electrons
(inner and outer) associated with it. Or, it can be
defined as the kernel charge minus the number of
outer electrons associated with it. Either calculation gives a formal charge of _____ for N
in the ammonia formula.

---

0 (+7 −2 −5 or +5 −5)

13-11 Suppose an oxygen atom $\overset{xx}{\underset{x}{x}}$ O $\overset{}{x}$ becomes attached to

the unbonded outer electron pair on nitrogen in ammonia:

$$
\begin{array}{c}
H \\
\overset{\cdot\cdot}{H:N:O}\overset{xx}{\underset{xx}{}}\overset{x}{\underset{x}{}} \\
H
\end{array}
$$

Splitting shared pairs gives

$$
\begin{array}{c}
H \\
H ·|· Ṅ ·|· Ö : \\
H
\end{array}
$$

The formal charge on hydrogen is _____ as before.

0

13-12   H   The atomic number of nitrogen is 7,
H ·|· Ṅ ·|· Ö :   its kernel charge is +5. The number
    H   of electrons in the immediate environ-
              ment of the nitrogen kernel in the
NH_3O structure is _____ .

4

13-13

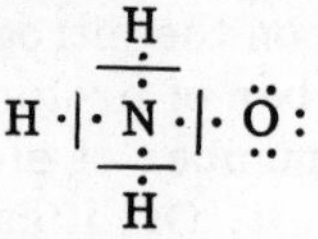

H · | · N̈ · | · Ö :   (with H above and H below the N)

With a kernel charge of +5 for N, and 4 electrons around it, the formal charge on nitrogen is _____. (Remember the sign.)

+1

13-14

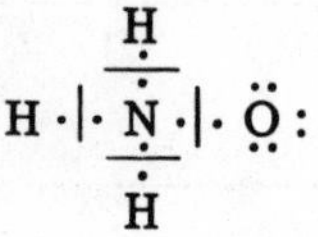

H · | · N̈ · | · Ö :   (with H above and H below the N)

The nitrogen in this compound is, therefore, positively charged (+1). Yet the molecule was made up of neutral ammonia and an atom of oxygen (neutral). Since the hydrogens are neutral (formal charge 0), we might expect the remaining atom, oxygen, to carry a formal charge of _____.

−1

13-15

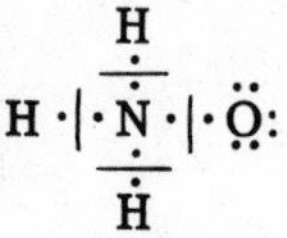

H · | · N̈ · | · Ö :   (with H above and H below the N)

The atomic number of oxygen is 8. It has _____ inner electrons. Its kernel charge, therefore, is

_____.

---

2   +6

13-16

$$H \cdot | \cdot \underset{\bullet}{\overset{\bullet}{N}} \cdot | \cdot \ddot{\ddot{O}} :$$

with H above and below the N.

The oxygen has a kernel charge of +6. In the immediate vicinity of the kernel are _____ electrons. Of these, _____ are not involved in bonding oxygen to other atoms ("unshared electrons"). The number of "shared" electrons associated with the oxygen atom above is _____.
The formal charge on oxygen (kernel charge —total outer electrons shared and unshared) must be

_____ .

---

7   6   1   −1   (i.e., +6 −7)

13-17 Our formula may then be written:

$$H : \overset{\bullet\bullet}{N} \overset{\oplus}{\underset{\bullet\bullet}{:}} \overset{\bullet\bullet}{O} \overset{\ominus}{:}$$

with H above and below the N.

It is a strongly polar molecule that should be strongly associated with neighboring molecules of the same formula. It would, therefore, be expected

to have a _____ boiling point than ammonia, since the latter has no formal charges.

---

higher

13-18 The nitrogen in ammonia has an unshared electron pair:

$$H : \overset{\bullet\bullet}{N} :$$

with H above and below the N.

The boron in boron trifluoride lacks an electron pair for a stable octet:

$$\begin{array}{c} \ddot{:}\ddot{F}\ddot{:} \\ \overset{\cdots}{:}\overset{\cdots}{F}:B \\ \ddot{:}\ddot{F}\ddot{:} \end{array}$$

In BF_3, if we split electron pair bonds as before, the boron kernel of charge +3 (i.e., +5 −2) has _____ electrons associated with it. The formal charge on boron then is _____.

---

3   0

13-19   If ammonia and boron trifluoride were to associate, both nitrogen and boron would have electron octets around them:

$$\begin{array}{ccc} H & & :\ddot{F}: \\ \overset{\cdots}{H}:\overset{\cdots}{N} & \dot{+} & B \dot{+} \ddot{F}: \\ H & & :\ddot{F}: \end{array}$$

In BF_3, boron had a formal charge of 0. Let us calculate what the formal charge will be in the formula above. Splitting shared bonds leaves _____ electrons associated with the boron kernel.

4

13-20
$$\begin{array}{ccc} H & :\ddot{F}: \\ \overset{\cdot}{H}\dot{+}\overset{\cdot}{N} \dot{+} B : \ddot{F}: \\ \overset{\cdot}{H} & :\ddot{F}: \end{array}$$
The formal charge on boron, with a kernel charge of +3, and 4 electrons around it, is _____. The nitrogen, kernel charge +5, and _____ electrons around it, has a formal charge _____. Since hydrogens again are neutral, and the whole molecule is neutral, we would expect the fluorines to have a formal charge of

_____.

−1   4   +1   0

13-21

$$H : \overset{\overset{\displaystyle ..}{..}}{\underset{\overset{\displaystyle ..}{H}}{N}} \overset{\oplus}{:} \overset{..}{\underset{\overset{\displaystyle ..}{F}}{B}} \overset{\ominus}{:} \overset{\overset{\displaystyle ..}{..}}{F} :$$

For any of the fluorine atoms, with atomic number 9 its kernel charge is _____. Splitting shared electrons, the number of electrons associated with the fluorine kernel is _____. The formal charge therefore is 0.

+7   7

13-22

$$H : \overset{\oplus}{\underset{\overset{\displaystyle ..}{H}}{N}} : \overset{..}{\underset{\overset{\displaystyle ..}{F}}{B}} \overset{\ominus}{:} \overset{..}{F} :$$

We may summarize: The formal charge is the kernel charge minus the number of unshared electrons minus half the number of shared electrons.

	Kernel Charge	No. Unshared e's	1/2 No. Shared e's	Formal Charge
N	+5	0	4	+1
B	+3	____	____	____

0   4   −1

13-23 In water $H : \overset{\overset{\displaystyle ..}{..}}{\underset{\displaystyle H}{O}} :$

	Kernel Charge	No. Unshared e's	1/2 No. Shared e's	Formal Charge
H	+1	0	1	0
O	+6	____	____	____

4   2   0

13-24   There are, therefore, no formal charges in the water molecule. (Its *polarity* is due to the differing electronegativities of H and O.) Suppose a proton is attached to an unshared electron pair of water to form the aqueous hydrogen ion H_3O^+; let us see whether the formal charge is on O or H   H : $\ddot{O}$: H
　　　　　　　　　　　　　　　　　　　　　　　　　　　　　　　　H

	Kernel Charge	No. Unshared e's	1/2 No. Shared e's	Formal Charge
H	+1	_____	_____	_____
O	+6	_____	_____	_____

	0	1	0
	2	3	+1

13-25   In H_3O^+, then, the positive charge resides on oxygen H : $\overset{\oplus}{\ddot{O}}$. H (except as modified by electroneg-
　　　　　　　　　H

ativity). The corresponding formal charges for the
　　　　　　　　　　　　　　　　　H
ammonium ion, NH_4^+, H : $\ddot{N}$: H are for N_____
　　　　　　　　　　　　　　　　　H

(atomic number 7), for H _____ .

+1   0

13-26   To convert the chlorine atom to a chloride ion, one electron is added:

　　　　: $\dot{\ddot{Cl}}$:　　　　　　　　　: $\ddot{\ddot{Cl}}$:

　　chlorine atom　　　　chloride ion, Cl^-

Chlorine has an inner shell of two electrons and a second complete shell of eight electrons. Its atomic number is 17. Its kernel charge, therefore, is

_____.

+7

13-27

              :Ċl: :C̈l:

             atom   ion

With a kernel charge of +7, the formal charge on the chlorine atom must be _____ and on the chloride ion _____.

---

0   −1

13-28 From the examples cited so far (e.g., Cl, Cl⁻, H_2O, H_3O^+, NH_3, NH_4^+, NH_3O, H_3NBF_3), you may have noticed that the sum of the formal charges for isolated atoms and neutral molecules always equals _____, while the sum for ions equals the charge on the ion. This is a general rule, made necessary by the construction from protons and electrons.

0

13-29 The chlorate ion may be written:

          :Ö:
          :Cl:Ö:
          :Ö:

With kernel charges +7 for Cl and +6 for O, the formal charges for Cl and for each oxygen must be _____ and _____ respectively.

+2   (Cl, +7 −2 −3)   −1   (O, +6 −6 −1)

13-30 With formal charges −1 on each oxygen and +2 on chlorine:

$$:\ddot{O}:^{\ominus}$$
$$:\ddot{C}l \textcircled{+2} \ddot{O}:^{\ominus}$$
$$:\ddot{O}:^{\ominus}$$

the over-all charge on the ion must be _____.

−1 (+2 −3)

13-31   With chlorate ion ClO_3^-, and sodium ion Na^+, the formula of sodium chlorate will be _____.

$NaClO_3$

13-32   The borohydride ion has the formula $H \div \overset{\overset{\textstyle H}{\bullet\bullet}}{\underset{\bullet\bullet}{B}} \div H$.

The formal charge on boron is _____ (atomic number of boron is 5). The formal charge on each hydrogen is _____. The charge on the BH_4 ion is, therefore, _____, and the formula of the lithium compound containing this ion (Li acts as Li^+) is

_____.

−1   0   −1 $LiBH_4$

13-33   An approximation to the sulfur dioxide formula is
:$\ddot{O}$:$\ddot{S}$::$\ddot{O}$:   Note that in this formula the sulfur and the
right-hand oxygen are bonded by four electrons.
When these electrons are divided equally, each atom
will gain the charges of two electrons. The formal
charges for the atoms from left to right are

_____, _____, and _____
(kernel charge of S = 6).

_____

        −1   +1   0

_____

13-34   To simplify formula writing, electron pairs are
often written as short single lines. Thus, H : H is

written H—H, and $\begin{array}{c} \text{H} \\ \text{H} : \ddot{C} : \text{H} \\ \ddot{\text{H}} \end{array}$ as $\begin{array}{c} \text{H} \\ | \\ \text{H}-\text{C}-\text{H} \\ | \\ \text{H} \end{array}$.

     Shared electron pairs are written as lines joining
the pair of bonded atoms. On this pattern, ethane,
$\begin{array}{c} \text{H} \quad \text{H} \\ \text{H} : \ddot{C} : \ddot{C} : \text{H} \\ \ddot{\text{H}} \quad \ddot{\text{H}} \end{array}$, can be written as

_____.

_____

$\begin{array}{c} \text{H} \quad \text{H} \\ | \quad\ | \\ \text{H}-\text{C}-\text{C}-\text{H} \\ | \quad\ | \\ \text{H} \quad \text{H} \end{array}$

_____

13-35   To distinguish *un*shared from shared electron pairs, unshared electron pairs are represented as lines *along the edge* of the atom kernel symbol— that is, *not* radiating out from the atom. Thus

$$H : \ddot{\underset{\cdot\cdot}{F}} : \quad = \quad H - \overline{\underline{F}} \,|$$

$$H : \underset{\overset{\cdot\cdot}{H}}{\ddot{N}} : H \quad = \quad H - \underset{\overset{|}{H}}{N} - H$$

$$H : \underset{\overset{\cdot\cdot}{H}}{\ddot{O}} : \quad =$$

_____.

$$H - \overline{\underset{\overset{|}{H}}{O}} \,|$$

13-36   How do we reformulate formal charge calculations in terms of these formulas? *Either* we can remember that each line represents two electrons or we can reformulate our earlier definition which read:

   Formal charge = kernel charge − No. of unshared electrons − 1/2 No. shared electrons

to read:

   Formal charge = kernel charge − 2 × No. of lines not shared − No. of lines (or single bonds) shared

The formal charge on carbon (atomic number 6) in the short-lived intermediate would be

_____.

+1

13-37   On the line formulation, nitrogen (N_2 or $:N:N:$

with dots above and below, or $:N:::N:$) would be written $|N{\equiv}N|$. Similarly, the carbon dioxide formula $:\ddot{O}::C::\ddot{O}:$ would be written _____.

$|\overline{O}{=}C{=}\overline{O}|$

13-38   Our earlier sulfur dioxide formula, written in the new form $|\overline{\underline{O}}{-}\overline{S}{=}\overline{O}|$ would have formal charges from left to right of _____, _____, _____.

−1   +1   0

13-39   One formula for the nitrite ion, NO_2^-, is $|\overline{\underline{O}}{-}\overline{N}{=}\overline{O}|$. The formal charges from left to right are _____, _____, _____.

−1   0   0

13-40   Conversion of nitrite to nitrate ion involves the addition of an oxygen atom to the unshared pair on nitrogen, thus

$$|\overline{\underline{O}}{-}N{\Big\langle}\begin{array}{c}|\overline{O}|\\ \\ \overline{O}|\end{array}$$

Formal charges: N _____
Each singly bound O _____
    doubly bound O _____
total charge _____

+1   −1   0   −1   [i.e., +1 + 2(−1)]

The Writing of Electronic Formulas

14-1   In Chapter 13 the formulas were supplied and you were asked only to calculate charges. How does one decide where the electrons should be placed? Where are the electrons in the perchlorate ion, ClO_4^-, in the sulfate ion, $SO_4^=$, in sulfuric acid, H_2SO_4? The procedure is first to determine the total number of electrons *available*—that is, those that are *not part of completed inner electron shells* (kernels). Water is composed of two hydrogens and an oxygen atom. The total number of *available* electrons is

_____.

---

8 (one from each H, six from O)

14-2   In ammonia the number of electrons outside of completed nonbonding shells supplied by one nitrogen and three hydrogen atoms is _____. (Atomic number of N is 7.)

---

8   $[5 + 3(1)]$

14-3   In nitric acid, HNO_3, the total number of available electrons is _____.

---

24   $[1 + 5 + 3(6)]$

14-4   In sulfuric acid, H_2SO_4, the total number of available electrons is _____ (atomic number of
S is 16, inner shells two and eight electrons).

---

32 $[2(1) + 6 + 4(6)]$

14-5   Knowing the number of electrons available for
bond formation, we need to know *which* atoms are
directly connected to each other. From elementary
chemical knowledge we can infer, for instance, that
since hydrogen practically never is bonded to more
than one atom, all hydrogens in ammonia must be
attached to nitrogen, giving the "atomic arrangement"

    H
  H N H . Given from chemical evidence that in nitric
acid, HNO_3, H is attached to O, and no oxygens are
attached to each other, the "atomic arrangement"

for nitric acid should be _____ .

          O
    H O N
          O

14-6   Analogously in sulfuric acid, H_2SO_4, chemical
evidence indicates that H is attached to O and no
oxygens are attached to each other. The atomic
arrangement for H_2SO_4 must, therefore, be

_____ .

        O   O
  H O S O (or H O S O H. We are not interested in
        O   O   bond direction at this
        H   stage.)

14-7   As a last example, the atomic arrangement for

COCl$_2$ must be _____, given that in this
molecule each chlorine is only bonded to one other
atom and oxygen is bonded only to carbon.

     Cl   O
O C   (or Cl C Cl, etc.)
     Cl

14-8   Given the atomic arrangement and the number of
electrons available, how do we distribute the latter?
In such a way that each atom, as far as possible,
has a noble gas configuration of electrons around it
—that is, two for hydrogen, eight in the outer shell
for elements Li to F, and also in most instances
from Na to Cl.

Shared electrons as $\substack{X\\X}$ in H $\substack{X\\X}$ Cl: belong to the
electron configurations of *both* atoms. Here H has
an electron configuration of _____ electrons
and Cl of _____ electrons.

---

2   8

14-9   In the aqueous hydrogen ion H :O: H, the electron
                                H
configuration around oxygen is one of _____
electrons.

8

14-10   We further place electrons in pairs, unless there is very strong chemical evidence against it (or if we have an odd number of electrons). All pairs of atoms attached to each other are connected at least by an electron pair bond, thus H O H becomes

                     H

H : O : H . *(Note that we are ignoring unshared elec-*

    H

trons and multiple bonds at this stage.) Similarly

           H   O

acetic acid, H C C   , becomes _____

           H   O H

if we follow the same procedure.

    H

          O

H : C̈ : C̈

    H   O : H

(We shall return to the discussion of acetic acid in Frame 14-30.)

14-11   The remaining available electrons are now distributed to give each atom a noble gas shell (i.e., two electrons for hydrogen, and usually eight for other atoms). In NH_3, six electrons are used in bonding, eight electrons are available. The final electronic structure therefore looks as follows:

_____.

     ..

H : N̈ : H

    H

14-12　　In HOCl, hypochlorous acid, there are 14 electrons available for bonding (H 1, O 6, Cl 7).

_____ are needed to bond the atoms by
　(number)
single electron pair bonds in the atomic arrangement

above. _____ further electrons are avail-
　　　　(number)
able to complete octets (or duets for hydrogen).

Four　　Ten

14-13　　H : O : Cl must then accommodate ten more electrons in a stable arrangement. This would look as follows: _____ .

H $:\ddot{\underset{\cdot\cdot}{O}}: \ddot{\underset{\cdot\cdot}{Cl}} :$

14-14　　What about a molecule such as carbon dioxide O C O? The number of electrons available for bonding is _____ (atomic numbers O 8, C 6).

16　[2(6) + 4]

14-15　　O C O　With 16 electrons available and _____ needed for bonding the two oxygens to carbon by

single covalent bonds, _____ electrons are left for octet formation.

4　　12

14-16   O : C : O   Where should the 12 electrons go?
Each oxygen needs _____ further electrons to
complete its octet and carbon needs _____ .

---

6   4

14-17   O : C : O   With each oxygen needing 6 more electrons and carbon needing 4, a total of 16 electrons
is still needed. But only 12 more are available.

_____ of these electrons must, therefore,
(number)
be *shared* so that they can belong to the octets of
both atoms.

Four   (i.e., 16 – 12)

14-18   O : C : O   Placing four more electrons in pairs
between bonded atoms leads to the structure(s)

_____ .

O : : C : : O   or   O : : : C : O

14-19   Note that there are two ways of distributing eight
electrons between the two C O bonds, O : : C : : O
or O : : : C : O. Taking first the O : : C : : O arrangement, each oxygen now *needs* _____ electrons
toward its stable octet and carbon *has* _____ .
(Since 8 electrons are still available, we should now
be able to complete the structure.)

---

4   8

14-20   O : : C : : O plus 8 electrons leads to the completed

structure _____.

:Ö : : C : : Ö :

14-21   Similarly O : : : C : O plus 8 electrons leads to

_____.

:O : : : C :Ö̤ :

14-22   Which structure |Ō=C=Ō| or |Ō—C≡O| is
likely to be closer to a correct description of the
carbon dioxide molecule? Let us calculate formal
charges.
For |Ō=C=Ō|

	Kernel Charge	No. Unshared e's	1/2 No. Shared e's	Formal Charge
O	+6	_____	_____	_____
C	+4	_____	_____	_____

+6	4	2	0
+4	0	4	0

14-23 A. $|\overline{O}{=}C{=}\overline{O}|$

	Kernel Charge	No. Unshared e's	1/2 No. Unshared e's	Formal Charge
O	+6	4	2	0
C	+4	0	4	0

B. $_{(1)}|\overline{\underline{O}}{-}C{\equiv}O|_{(2)}$

$O_{(1)}$	+6	___	___	___
$O_{(2)}$	___	___	___	___
C	___	___	___	___

$O_{(1)}$	+6	6	1	−1
$O_{(2)}$	+6	2	3	+1
C	+4	0	4	0

Note: (a) $|\overline{O}{=}C{=}\overline{O}|$ (b) $|\overset{\ominus}{\underline{\overline{O}}}{-}C{\equiv}\overset{\oplus}{O}|$

We may ask which of the two structures (a) or (b) is likely to be a more accurate description of the CO_2 molecule. From elementary electrostatic considerations, we know that energy is required to keep opposite charges apart. Since (b) requires such charge separation, it will be the less probable form. Yet, since both forms are permitted by our rules for molecule construction, the electrons are likely to distribute themselves in an arrangement some-where between (a) and (b). The discussion of such problems belongs to the *Theory of Resonance* (see Chapter 15).

14-24 Working on a simpler case, that of hydrogen cyanide, HCN, the number of *available* electrons is

_____ .

10 (H 1, C 4, N 5)

14-25   Single bonding leads to H : C : N, leaving _____ electrons to be assigned.

6

14-26   In H : C : N, C needs _____ more electrons to complete its octet, and nitrogen needs

_____ , making a total of _____ .

---

4   6   10

14-27   H : C : N   With 10 electrons needed and only 6 more available, _____ must be involved in

further bond formation. Since H already has the helium configuration, the 4 electrons must bond C and N. The HCN formula must then look like this:

_____ .

---

4   H : C : : : N

14-28   H : C : : : N or H—C≡N
We have now assigned 8 of our 10 electrons. The remaining 2 must be used to complete an octet

giving the electronic and line formulas _____

and _____ .

H : C : : : N :   H—C≡N|

14-29   H—C≡N|

Formal charge calculations lead to formal charges from left to right of _____, _____, _____.

---

0   0   0

14-30   In Frame 14-10 we arrived at a partial formulation for acetic acid as

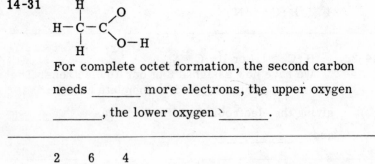

The *total* number of *available* electrons is _____, Since 14 of these are already assigned, we still need to dispose of _____.

---

24   10

14-31

H—C—C⟨O O—H   (with H above and below the first C)

For complete octet formation, the second carbon needs _____ more electrons, the upper oxygen _____, the lower oxygen _____.

---

2   6   4

14-32

```
       H      O
       |     //
  H — C — C
       |     \
       H      O — H
```

With 12 more electrons needed and only 10 more available, 2 must be shared. Assigning these 2 electrons, leads to *two* possible structures for acetic acid.

_____   _____

```
       H      O                    H      O
       |     //                    |     //
  H — C — C                   H — C — C
       |     \                      |      \\
       H      O — H                 H       O — H
```

14-33

```
       H      O                        H      O
       |     //                        |     //
  H — C — C                       H — C — C
       |     \                          |      \\
       H      O — H                     H       O — H
```

The remaining 8 electrons are now assigned in such a way as to complete octets, yielding

_____ and _____.

```
       H      ⟨O⟩                       H      ⟨O⟩
       |     //                         |     //
  H — C — C                        H — C — C
       |     \                           |      \\
       H      O̱ — H                      H       O̱ — H
```

14-34   Calculation of formal charges for

H   O   H   O
| ⫽ | ⫽
H—C—C   H—C—C
| ＼ | ＼
H   O—H   H   O—H

gives final formulas, with formal charges indicated
at atoms carrying them:

_____   _____

H   O   H   O⊖
| ⫽ | ⫽
H—C—C   H—C—C
| ＼ | ＼
H   O—H   H   O—H
                                    ⊕

Note: We will return to acetic acid in the next chapter.

14-35   Nitric acid, HNO_3, is bonded, as a first approxima-

                         O
                         ‖
tion, as follows: H—O—N   . With a total of _____
                         |
                         O

_____ available electrons, and 8 utilized as

above, _____ are left for octet completion.
The number of electrons still needed for octet formation in the absence of further bonding is

_____ . The number of further electrons involved

in bonding is, therefore, _____ .

24   16   18   2 (i.e., 18 − 16)

14-36   There are several positions for the further electron pair, leading to the formulas

_____ , _____ , and _____ .

H—O—N(=O)(—O)   H—O—N(=O)(—O)   H—O=N(=O)(—O)

$$\text{H}-\text{O}-\text{N}\begin{smallmatrix}\nearrow\text{O}\\\searrow\text{O}\end{smallmatrix} \qquad \text{H}-\text{O}-\text{N}\begin{smallmatrix}\nearrow\text{O}\\\searrow\text{O}\end{smallmatrix} \qquad \text{H}-\text{O}=\text{N}\begin{smallmatrix}\nearrow\text{O}\\\searrow\text{O}\end{smallmatrix}$$

14-37   With 14 more electrons to be assigned in order
to complete octets, the formulas

$$\text{H}-\text{O}-\text{N}\begin{smallmatrix}\nearrow\text{O}\\\searrow\text{O}\end{smallmatrix} \quad , \quad \text{H}-\text{O}-\text{N}\begin{smallmatrix}\nearrow\text{O}\\\searrow\text{O}\end{smallmatrix} \quad ,\text{ and } \text{H}-\text{O}=\text{N}\begin{smallmatrix}\nearrow\text{O}\\\searrow\text{O}\end{smallmatrix}$$

become

_____ , _____ , and _____ .

$$\text{H}-\overline{\underline{\text{O}}}-\text{N}\begin{smallmatrix}\nearrow\overline{\text{O}}|\\\searrow\underline{\text{O}}|\end{smallmatrix} \qquad \text{H}-\overline{\underline{\text{O}}}-\text{N}\begin{smallmatrix}\nearrow\text{O}\rangle\\\searrow\text{O}\rangle\end{smallmatrix} \qquad \text{H}-\overline{\text{O}}=\text{N}\begin{smallmatrix}\nearrow\text{O}\rangle\\\searrow\text{O}\rangle\end{smallmatrix}$$

14-38   When we complete the formulas by assigning formal charges, the formulas

$$H-\overline{O}-N\begin{matrix} \overline{O}| \\ \\ \underline{O}\rangle \end{matrix}\ ,\quad H-\overline{O}-N\begin{matrix} \langle\overline{O}\rangle \\ \\ \underline{O}\rangle \end{matrix}\ ,\ \text{and}\ H-\overline{O}=N\begin{matrix} \langle\overline{O}\rangle \\ \\ \langle\underline{O}\rangle \end{matrix}$$

become

_____ , _____ , and _____ .

$$H-\overline{O}-\overset{\oplus}{N}\begin{matrix} \overline{O}| \\ \\ \underline{O}|^{\ominus} \end{matrix}\quad H-\overline{O}-\overset{\oplus}{N}\begin{matrix} \langle\overline{O}\rangle^{\ominus} \\ \\ \underline{O}\rangle \end{matrix}\quad H-\overset{\oplus}{O}=\overset{\oplus}{N}\begin{matrix} \overline{O}|^{\ominus} \\ \\ \overline{\underline{O}}|^{\ominus} \end{matrix}$$

14-39

$$H-\overline{O}-\overset{\oplus}{N}\begin{matrix} \overline{O}| \\ \\ \underline{O}|_{\ominus} \end{matrix}\qquad H-\overline{O}-\overset{\oplus}{N}\begin{matrix} \langle\overline{O}\rangle^{\ominus} \\ \\ \underline{O}\rangle \end{matrix}\qquad H-\overset{\oplus}{O}=\overset{\oplus}{N}\begin{matrix} \overline{O}|^{\ominus} \\ \\ \overline{\underline{O}}|_{\ominus} \end{matrix}$$

   (a)   (b)   (c)

Since the third formula has twice as many separated charges as the first two and in addition has positive charges on adjacent atoms, it will be a much less stable structure than the first two. Again, resonance theory must be invoked to decide how (a) and (b) determine the actual structure.

The complete formula with formal charges for nitrous acid, HONO, is _____ .

$$H-\overline{O}-\overline{N}=\overset{\,}{O}\rangle\qquad \text{or}\qquad H-\overset{\oplus}{\underline{O}}=\overline{N}-\overset{\ominus}{\underline{O}}|$$

14-40 In applying our formula-writing procedure to ions, we must remember that the ionic charge implies an excess or deficiency of electrons in the ion as compared with the original collection of atoms. Thus Cl^- has 8 outer electrons as against the chlorine atom's 7. The nitrate ion, NO_3^-, has _____ available electrons, even though one nitrogen atom and three oxygen atoms together only supply

_____ .

---

24   23

14-41 If we look at the perchlorate ion, ClO_4^-, the chlorine *atom* brings 7 electrons, four oxygens a total of 24, giving a total of 31, and the negative charge implies the presence of one further electron (from some atom that became a cation). Thirty-two electrons distributed around

       O
O Cl O   leads to
       O

_____ .

$$|\overline{\underline{O}}|$$
$$|\overline{\underline{O}} - \overset{|}{\underset{|}{Cl}} - \overline{\underline{O}}|$$
$$|\underline{O}|$$

14-42   Formal charges for $|\overline{\underline{O}}-\overset{\displaystyle |\overline{\underline{O}}|}{\underset{\displaystyle |\underline{O}|}{Cl}}-\overline{\underline{O}}|$ leads to the

formula _____ .

$$\overset{\ominus}{|\overline{\underline{O}}}-\overset{\displaystyle |\overline{\underline{O}}|^{\ominus}}{\underset{\displaystyle |\underline{O}|^{\ominus}}{Cl}}\overset{+3}{}-\overline{\underline{O}}|^{\ominus}$$

14-43   The chlorate ion, $ClO_3^{\ominus}$ (again all oxygens attached only to Cl), showing all outer electrons and
formal charges, becomes

_____ .

$$\overset{\ominus}{|\overline{\underline{O}}}-\overset{\displaystyle |\overline{\underline{O}}|^{\ominus}}{\underset{\displaystyle |\underline{O}|_{\ominus}}{Cl|}}\overset{\ominus 2}{}$$

14-44   The sulfate ion, $SO_4^{=}$, has _____ electrons
to be distributed.

32

14-45   With all oxygens attached to sulfur, the complete
formula for $SO_4^=$, including formal charges, must be

_____ .

_____

$$^\ominus|\underline{O}-\overset{\displaystyle |\overline{O}|\,^\ominus}{\underset{\displaystyle |\underline{O}|\,^\ominus}{\overset{|}{\underset{|}{S}}\!\!\!\overset{+2}{}}}-\overline{O}|\,^\ominus$$

Note that the sum of the formal charges in

$$^\ominus|\underline{O}-\overset{\displaystyle |\overline{O}|\,^\ominus}{\underset{\displaystyle |\underline{O}|_\ominus}{\overset{|}{\underset{|}{S}}\!\!\!\overset{+2}{}}}-\overline{O}|\,^\ominus$$

is −2, as it must be since sulfate is a dinegative ion. Note
also that since sulfur is in the second period of eight in the
periodic table, it can expand its valence shell beyond 8
electrons. It is likely, therefore, that the following structures also are permissible:

$$_\ominus|\underline{O}-\overset{\displaystyle ^\ominus|\overline{O}|}{\underset{\displaystyle |\underline{O}|\,_\ominus}{\overset{|}{\underset{|}{S}}\!\!\!\overset{\oplus}{}}}=O\rangle \qquad\qquad ^\ominus|\underline{O}-\overset{\displaystyle \overline{O}|}{\underset{\displaystyle _\ominus|\underline{O}|}{\overset{\|}{\underset{|}{S}}}}=O\rangle$$

14-46   Ignoring the possibility of valence shell expansion, the completed formula for sulfuric acid (H's attached to O, O's not attached to each other) including formal charges is

_____ .

14-47   The complete formula for phosgene, $COCl_2$, (Cl's and O attached only to carbon) may be

_____ or _____ or _____ .

There are, of course, many compounds with insufficient electrons, or other good reasons, for not obeying the octet rule. These compounds have been the agents keeping valence theorists busy. The boron hydrides are good examples. The octet rule, like most scientific generalizations, is an oversimplification. It is valid for almost all organic compounds and numerous inorganic ones. It is, therefore, an enormously helpful rule, even as a first step to the understanding of the exceptions.

The Writing of Resonance Formulas

In the chapter on the writing of electronic formulas we discovered that for carbon dioxide, CO_2, the available electrons could be distributed in more than one way so that each atom was surrounded by 8 electrons. The three possible electron distributions around a given set of atoms O C O are as follows:

$$\langle O = C = O \rangle \qquad |\overset{\ominus}{\underline{O}} - C \equiv O| ^{\oplus} \qquad |O \equiv C - \underline{O}|^{\ominus}$$

or by the electron dot formulation:

$$: \overset{..}{O} : : C : : \overset{..}{O} : \qquad : \overset{..}{\underset{..}{O}}{}^{\ominus} C : : : \overset{..}{O}{}^{\oplus} \qquad : \overset{\oplus}{O} : : : C : \overset{..}{\underset{..}{O}}{}^{\ominus}$$

$$\quad A \qquad\qquad\qquad B \qquad\qquad\qquad C$$

Since each atom retains an octet of electrons around it, the electrons are not likely to be held rigidly in just one of the three distributions. Instead they will be free to move within the limits set by the three forms. The electrons are likely to spend most of the time somewhere between these positions. However, the three electron distributions are not equally probable. Distributions B and C involve the creation of $\oplus$ and $\ominus$ charges. Since these opposite charges tend to neutralize each other (and this can be done by a simple shift of electrons back to distribution A), we may suspect that the electrons of CO_2 at room temperature will remain closer to distribution A than to B and C.

If A, B and C were equally probable distributions, we would represent the actual CO_2 molecule as a superposition of the three formulas with each one considered equally important. Such a composite representation might resemble the following:

$$\overset{\frac{1}{3}\oplus}{|\underline{O}} \equiv\equiv\equiv C \equiv\equiv\equiv \overset{\frac{1}{3}\oplus}{\underline{O}|}$$

In each of the three formulas A, B and C the central carbon is neutral. It is, therefore, neutral also in the combined formula. Each oxygen is $\ominus$ in one-third of the formulas and $\oplus$ in one-third. And when one O is $\oplus$ the other is $\ominus$. This is represented by the formal charges indicated on the composite formula. Finally, the C—O bonds must be somewhere between single and triple.

However, we have said that, in fact, the three carbon dioxide structures are probably not equally significant. The charges on the oxygens should, therefore, be indicated as less than one-third of a unit charge. We use the Greek letter δ (delta) to represent a fraction of a unit charge of unspecified amount and write carbon dioxide as

$$\overset{\delta\oplus}{O} \equiv\equiv C \equiv\equiv \overset{\delta\oplus}{O}$$

15-1   Carbon dioxide is a quite complicated example of the resonance phenomenon so let us begin with something much simpler. Acetic acid ionizes in water to the acetate ion:

$$CH_3CO_2H + H_2O \rightleftarrows CH_3CO_2^- + H_3O^+$$

An electronic formula for the acetate ion that has 8 electrons around each carbon and oxygen atom can be represented as

_____.

The ⊖ charge of the acetate ion is situated on one

_____ atom.
(element)

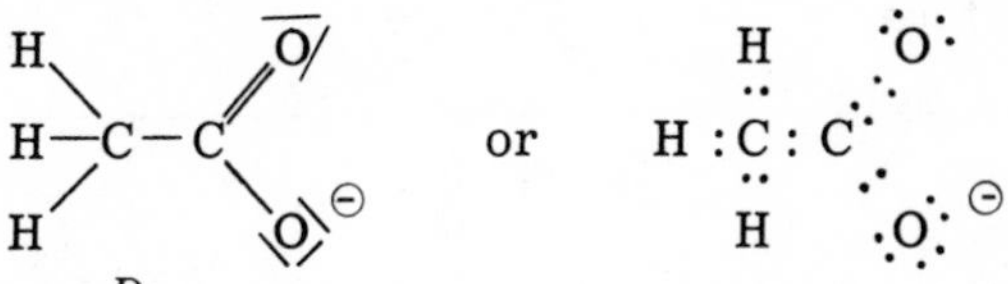

D

The ⊖ charge is on one *oxygen* atom.

15-2   In the formula of the acetate ion one oxygen is doubly bound, the other singly. But we could just as well have written the formula with the bonds, un-shared electrons, and formal charges on the two oxygens, interchanged. Using the line notation (as in D above) the two ways of writing the acetate ion are

_____ and _____ .

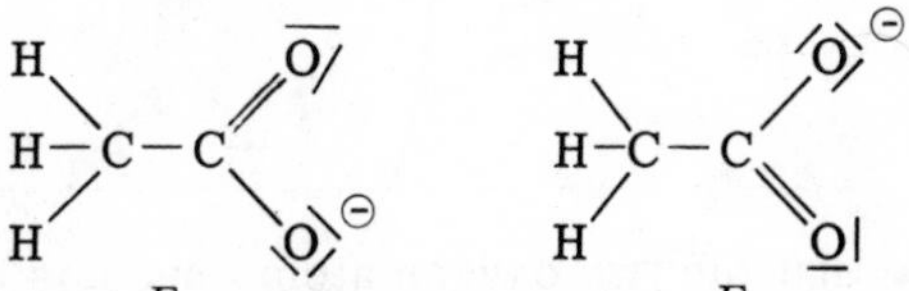

E F

15-3   Now these two structures (are/are not) of equal energy.

are of equal energy

15-4   Therefore, we can give equal weight to E and F in representing the actual acetate ion. A given oxygen carries one unit negative charge in one formula and is neutral in the other. The fractional charge assigned to each oxygen in the composite formula will, therefore, be _____.

1/2

15-5   Similarly, since the C—O bond is single in one formula and double in the other, each C—O bond can be thought of as a "$1\frac{1}{2}$ bond," and we will write it as a single plus a partial bond, thus C≡O.
    The actual structure of the acetate ion may then

be represented as

$$H-C\begin{matrix} H \\ | \\ H \end{matrix}$$

_____ (complete the formula showing bonds and charges)

$$H-\overset{\displaystyle H}{\underset{\displaystyle H}{C}}-C\overset{O^{\frac{1}{2}\ominus}}{\underset{O^{\frac{1}{2}\ominus}}{}}$$

15-6   Notice that the two oxygen atoms become completely equivalent in the acetate ion according to this treatment. Spectroscopic information supports this conclusion.
    Suppose we look at a very similar structure—in fact an isoelectronic one—the molecule nitromethane, CH_3NO_2.

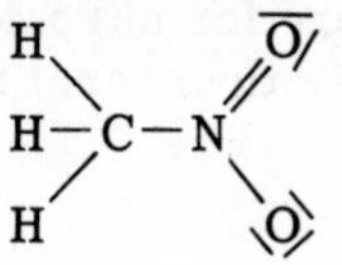

H ̄O ̄/
H—C—N
H \O/

When formal charges are inserted in this formula, it becomes:

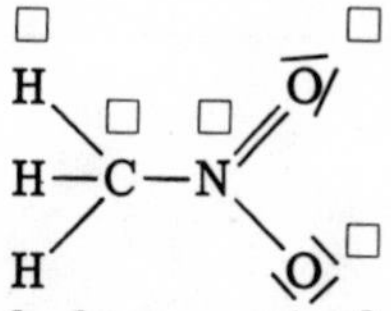

☐ ☐
H ☐ ☐ ̄O ̄/
H—C—N
H \O/ ☐

(insert formal charges —including zero—in boxes)

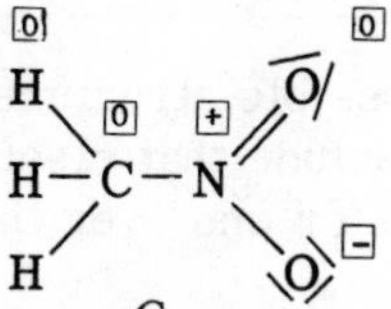

[0] [0]
H [0] [+] O/
H—C—N
H \O/ [−]
    G

15-7   Are there alternate distributions of the electrons in this formula, as there were in the case of the

acetate ion? _____

---

yes

15-8   A second structure for nitromethane (including
formal charges other than zero) is

_____.

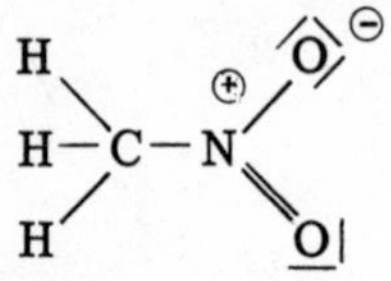

H

15-9   When we look at the two possible structures for
nitromethane, *G* and *H*, we conclude that electrons
(have no reason to/should) prefer one over the other.

---

have no reason to

15-10   The actual structure may, therefore, be represented as an equal superposition of the two structures
G and *H*. Since the nitrogen atom has a "plus one"
charge in each case, it will have a _____
charge in the combined representation.

+1   (The electrical state of the nitrogen atom is un-
affected by the redistribution of electrons.)

15-11   Since in the case of either oxygen, one formula assigns it a _____ charge and the other a _____ charge, the charge to be assigned to each oxygen on the composite formula will be _____.

−1   0   −$\frac{1}{2}$

15-12 The combined representation for nitromethane is, therefore,

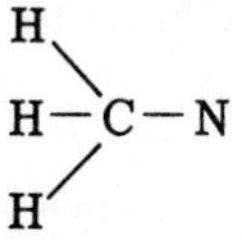

H
\
H—C—N
/
H   _____ (complete the structure)

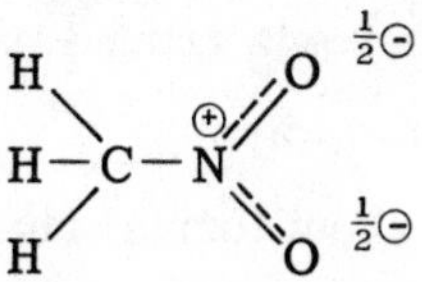

H
\
H—C—N$^{\oplus}$
/   O $^{\frac{1}{2}\ominus}$
H   O $^{\frac{1}{2}\ominus}$

15-13   Let us look at another case. There are two classical representations for benzene, C_6H_6, a six-carbon ring structure of alternating single and double bonds. From this description the two structural formulas

must be _____ and _____.

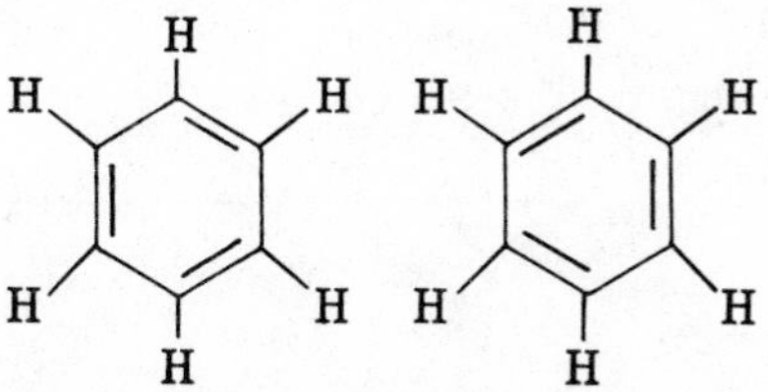

where each corner of the hexagon is occupied by a carbon atom.

15-14   Since no formal charges appear in either structure, none will appear in the combined representation

which may be written _____.

---

benzene

This structure is sometimes written

or

15-15   The sulfur dioxide molecule (O S O) requires the
distribution of 18 valence electrons around the three

atoms. This can be done in two ways: _____

and _____. (Forget about formal charges
for the moment.)

$|\overline{\text{O}}-\overline{\text{S}}=\text{O}\rangle$ and $\langle\text{O}=\overline{\text{S}}-\overline{\text{O}}|$

15-16   On inserting formal charges these formulas be-

come _____ and _____.

$|\overset{\ominus}{\overline{\text{O}}}-\overset{\oplus}{\overline{\text{S}}}=\text{O}\rangle$ and $\langle\text{O}=\overset{\oplus}{\underline{\text{S}}}-\overset{\ominus}{\underline{\text{O}}}|$

15-17   Again the two formulas are of equal energy and we say accordingly that they "contribute equally" to the actual structure of the molecule which may be

written by the composite structure _____.

$$\overset{\frac{1}{2}\ominus}{\underline{|\underline{O}|}}=\overset{\oplus}{\underline{S}}=\overset{\frac{1}{2}\ominus}{\underline{\underline{O}|}}$$

15-18   Our approach would predict that the two oxygens are equivalent, and that the molecule has a plane of symmetry through the sulfur atom. Charge cloud considerations would further predict that the mole-

cule is bent $O^{\diagup\overset{\overset{..}{S}}{\diagdown}}O$ because the unshared electron

pair on sulfur requires space, thus pushing the two oxygen atoms out of the linear O—S—O arrangement. Spectroscopic data confirm these conclusions.

When more than one structure is written for a molecule or ion and each satisfies the octet rule (or some further extension of valence theory), we speak of the separate structures as *resonance structures* and sometimes call the combined representation a *resonance hybrid.*

We usually write acetic acid as $H-\overset{\overset{\displaystyle H}{|}}{\underset{\underset{\displaystyle H}{|}}{C}}-\overset{\overset{\displaystyle /O\backslash}{\|}}{C}-\underline{\underline{O}}-H$,

I

but there is a second resonance structure that can

be written for it, _____ (include formal charges), as we saw in the previous chapter.

$H-\overset{\overset{\displaystyle H}{\diagdown}}{\underset{\underset{\displaystyle H}{\diagup}}{C}}-C\overset{\diagup O\backslash^{\ominus}}{\underset{\overset{\displaystyle \oplus}{\underline{O}}-H}{\diagdown}}$

J

15-19   The lower energy structure of the two structures
I and *J* is _____ because _____

_____ .

I, because *J* involves creation of opposite charges.
Since these tend to neutralize each other, electrons
tend toward structure *I*.

15-20   The two resonance structures, therefore, should
not be given equal weight. The actual structure will
be closer to *I*. We will therefore use δ+, δ− instead
of ½+, ½− in representing over-all formal charges.
The resonance hybrid formula, therefore, is represented as

_____ .

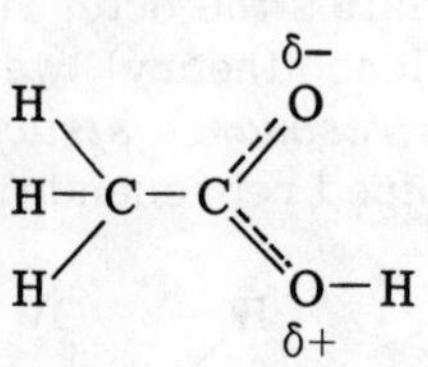

15-21   In the case of acetamide, $CH_3\overset{\displaystyle O}{\overset{\|}{C}}NH_2$, the two res-
onance structures are

_____ and _____ .

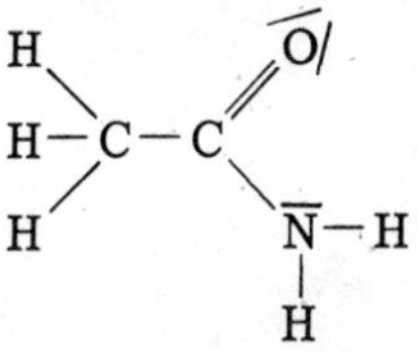

and

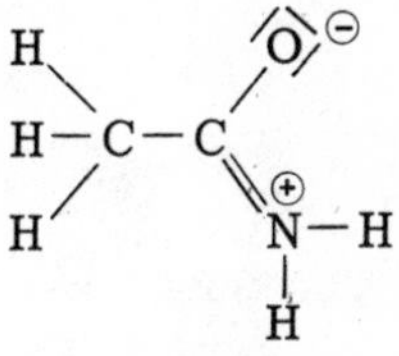

15-22   And the resonance hybrid structure for acetamide

is _____ .

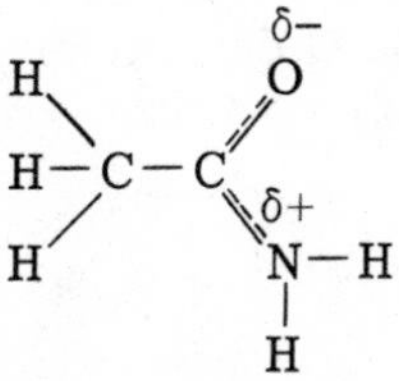

15-23   For practice write all resonance forms for the nitrate ion, NO_3^-, given that one of them is:

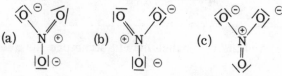 and then write the hybrid structure

The three resonance forms are

(a)   (b)   (c)

and the resonance hybrid is

Note that charge on $O = 2^{\ominus}$ divided by 3 atoms of oxygen. Hence net charge on ion $= 3\,(-2/3) + 1 = -1$

15-24   Now repeat this process for the carbonate ion,

$CO_3^=$ _____, and the nitrite ion,

NO_2^-, _____.

carbonate ion:

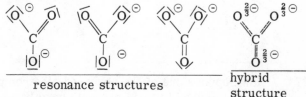

resonance structures

hybrid structure

nitrite ion:

resonance structures: ⟨O=N̄—O̲|⊖ and ⊖|O̲—N̄=O⟩

hybrid structure: O=N̄=O
                  ½⊖   ½⊖

15-25   Finally, let us look at a number of benzene derivatives. If an **OH** group replaces H in benzene we obtain the acidic substance, phenol, C_6H_5OH. Five resonance structures can be written. The first has no formal charges and is represented as

_____.

(or its mirror image)

15-26 The second structure merely exchanges double
and single bonds in the ring, giving

_____ .

15-27 Now suppose one of the unshared electron pairs
on oxygen is used to form a double bond between
oxygen and the adjacent ring carbon. This would
put 10 electrons around the ring carbon, so one electron pair has to become unshared. Since the molecule
must not come apart, the only electron pair which the
carbon can relinquish must be the one belonging to
the double bond between carbon atoms 1 and 2. If
this pair becomes an unshared pair on carbon atom
number 2, the resulting resonance structure is

_____ . (Use formal charges where
appropriate.)

unshared electron pair

15-28   The fourth resonance structure is the mirror image of the one just written

_____ .

15-29   Now, if the unshared electron pair on carbon atom 6 is shared between atoms 6 and 5, and the second electron pair of the double bond 4—5 is relocated to fall on atom 4, we obtain our last structure

_____ .

15-30   Combining these five structures we obtain our resonance hybrid structure

_____ .

You should convince yourself that, by the rules we have used, there is no way of putting ⊖ charges on carbon atoms 3 and 5.

15-31   The ammonia derivative of benzene is aniline, $C_6H_5NH_2$. Again five resonance forms can be written

of which the first is

The others are:

_____ , _____ , _____ , and _____ .

(Note these formulas do not show the ring hydrogens nor the N—H bonds.)

15-32   The hybrid structure of aniline then is

_____ .

15-33   Phenol as we have said can act as an acid. It ionizes in water to give the phenolate (or phenoxide) ion, $C_6H_5O^-$. The resonance structures for this ion are:

_____ :

15-34 The hybrid structure of the phenoxide ion becomes

_____ .

15-35   Naphthalene in one of its resonance forms can be written

Other resonance forms are _____ .

15-36 The hybrid naphthalene structure is represented as

_____.

  or sometimes as

15-37   One resonance form of anthracene can be written

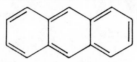

Its other resonance forms are

_____.

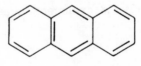

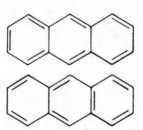

15-38 Phenanthrene can be written ,

but also as _____.

15-39 1,3-cyclopentadiene has the structure

_____.

15-40   If one of the H atoms of the CH_2 group is ionized as a proton, we are left with a $C_5H_5^{\ominus}$ ion. Its structure is

_____ .

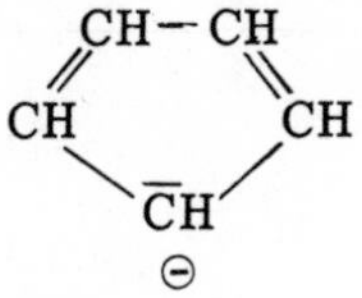

15-41   But this is not the only possible structure. There are several others:

_____ .

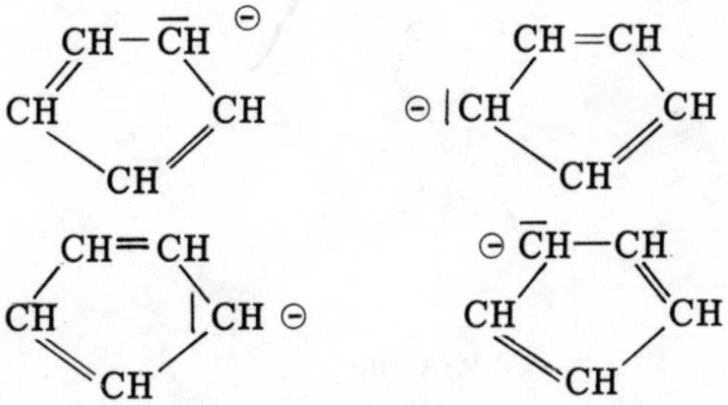

15-42   These five structures (are/are not) all equivalent.

---

are equivalent

15-43   The resonance hybrid of the five structures above, therefore, has the structure (including computed charge on each carbon; do not use δ)

_____ .

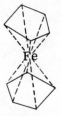

15-44   This is quite a remarkable ion. And it forms remarkable compounds such as $(C_5H_5)_2Fe$, where an iron (II) ion is sandwiched in between two slices of cyclopentadienyl anion:

Ferrocene

Isomeric Alcohols —
Their Number and Structures

Seventeen different six-carbon alcohols can be written by different skeleton formulas and can be named by the IUPAC system. They have all been prepared in the laboratory. How can we write them, with some certainty that we have not written duplicate structures? Is there a method other than trial and error for writing all the structural formulas corresponding to a given molecular formula? For substances of general formula $R-Y$ the answer is yes, if R is an alkyl group and Y is some functional group such as halogen, OH, NO_2, etc. attached to the alkyl group by a single bond. When $R = C_6H_{13}$ and $Y = OH$ our statement above can be rephrased to state that there are seventeen different ways of attaching six carbons, thirteen hydrogens and an OH group to each other, each corresponding to a distinct skeleton structure.

The method here presented is based on two papers by H. R. Henze and C. M. Blair [*J. Amer. Chem. Soc.* **53**, 3044, 3077 (1931); cf. also D. Perry, *ibid* **54**, 2918 (1932)]. We proceed from the simplest alcohol and work our way up to the six-carbon alcohols.

16-1   With one carbon we have methanol, CH_3OH.
       With two carbons, we have _____ of

       formula _____.

---

       ethanol C_2H_5OH

16-2   With 3 carbons the OH group can be either at the end or in the middle of the carbon chain, i.e., $C-C-C-OH$ or $C-C-C$. Let us isolate the

$$C-C-C$$
$$|$$
$$OH$$

$C-OH$ group which every alcohol must contain. There will be three valences to be filled at the carbon atom attached to OH. In the case of the three-carbon alcohols, two carbons need to be assigned to the valences of the $C-OH$ group. Both carbons can be attached to the *same* valence in the form of an

ethyl group, or _____.

---

or the two carbons can be attached to different valences as two methyl groups.

$$\begin{array}{c} C \\ \diagdown \\ C-OH \\ \diagup \\ C \end{array}$$

16-3   Two 3-carbon alcohols therefore exist. (We named them earlier as *n*- and isopropyl alcohol.) What about four-carbon alcohols? The number of different ways in which 3 carbons can attach themselves to the $-\overset{|}{\underset{|}{C}}OH$ group is _____.

4

16-4   The answer may puzzle you. It is derived as follows. Either all three carbons are attached to one of the $-\overset{|}{\underset{|}{C}}OH$ valences; or two carbons are attached to one valence and one to a second valence; or the three carbons are each attached to a separate valence. The three possibilities can be represented schematically in a table, each column listing the number of carbon atoms attached to a given valence. If the three valences are designated as x, y, and z, our table for the four-carbon alcohols appears as follows:

Number of carbons attached to valences of $\overset{\diagdown}{\underset{\diagup}{-}}$COH group:

	x	y	z
Case A	3	0	0
Case B	2	1	0
Case C	1	1	1

$$\begin{matrix} x \diagdown \\ y - C - OH \\ z \diagup \end{matrix}$$

Thus, the three carbons can attach themselves either as a propyl group; or as an ethyl plus a methyl group; or finally as three methyl groups. But we have said that there were four different ways of attaching the three carbons. The discrepancy lies in the fact that there are two ways in which a propyl group can attach itself to a valence link (see Frame 16-3 above). Line A in the table therefore corresponds to two different 4-carbon alcohols while lines B and C correspond to one isomer each. Four butyl alcohols with different skeletons, therefore,

exist. They have the formulas _____, _____,

_____, and _____.

C — C — C — C — OH   C — C — C — OH
                              |
                              C

C — C
       \
         C — OH
       /
C

C
  \
C — C — OH
  /
C

16-5   What about 5-carbon alcohols? We need to distribute four carbons over one, two, or three bonds of

the $\overset{\diagdown}{\underset{\diagup}{C}}$—OH carbon. The four carbons may be dis-

tributed over the three bonds as follows:

	x	y	z
A	4	0	0
or B	____	____	____ (Fill in the blanks.)
or C	____	____	____
or D	____	____	____

B	3	1	0
C	2	2	0
D	2	1	1

16-6   The number of 5-carbon alcohols having the
2-1-1 distribution (D) is _____ with formula(s)   .

_____.

1   C—C
               \
       C—C—OH
       /
       C

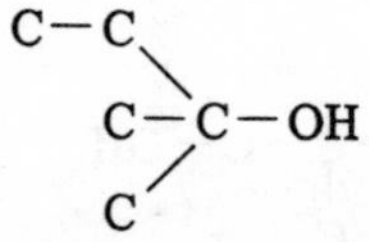

16-7   The number of 5-carbon alcohols having the 2-2-0

distribution (C) is _____ with formula(s)_____ .

1
```
C—C
    \
     C—OH
    /
C—C
```

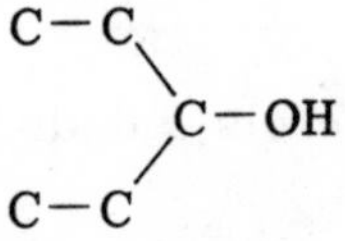

16-8   The number of 5-carbon alcohols having the 3-1-0
distribution (B) is _____ .

---

2   (See Frames 16-4 and 16-3.)

16-9   The formulas of the two 5-carbon alcohols with
the 3-1-0 distribution around the C—OH group are

_____ and _____ .

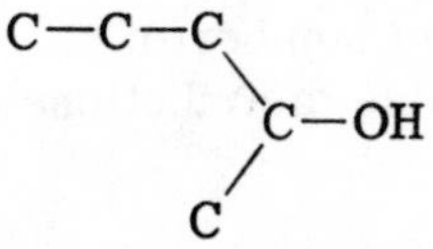

```
C—C—C
      \
       C—OH
      /
     C
```
and
```
     C
      \
       C
      / \
     C   C—OH
      \
       C
```

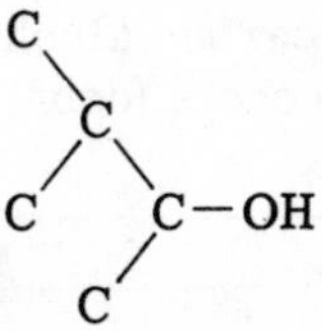

16-10   Now we must seek the number of 5-carbon alcohols with the distribution 4-0-0 (A). This is equal to
the total number of four-carbon alcohols (see Frame
16-4) because the number of different 4-carbon alcohols equals the number of different ways by which

a group of four linked carbon atoms can be attached by a single bond to *any* atom. The number we are seeking is, therefore, _____.

4

16-11   The four 5-carbon alcohols with distribution 4-0-0 have the formulas

_____, _____, _____, and _____.

```
C—C—C—C              C
           \          \
            C—OH       C—C
                      /     \
                     C       C—OH

C—C—C                 C
     |  \              \
     C   C—OH   and     C—C
                       /   \
                      C     C—OH
```

16-12   This completes the possible ways of constructing 5-carbon alcohols. The *total* number of 5-carbon alcohols (counting all possible distributions) is

_____.

8:   4(4-0-0) + 2(3-1-0) + 1(2-2-0) + 1(2-1-1) = 8

16-13   At last we come to the 6-carbon alcohols. We take off the COH group as usual. Now we study the distribution of the remaining five carbons over the three available bonds:

x	y	z
5	0	0
___	___	___

(You fill in the rest.)

___	___	___
___	___	___
___	___	___

4	1	0
3	2	0
3	1	1
2	2	1

16-14   The number of 6-carbon alcohols with the distribution 5-0-0 is _____.

8   (the same as the total number of 5-carbon alcohols (see Frame 16-12)

16-15   The number of 6-carbon alcohols with the distribution 4-1-0 is _____.

4   (see Frame 16-4)

16-16   The number of 6-carbon alcohols with the distribution 3-2-0 is _____.

The number of 6-carbon alcohols with the distribution 3-1-1 is _____.

The number of 6-carbon alcohols with the distribution 2-2-1 is _____.

3-2-0

2:   C—C—C
          C—C—C—OH
and
C
    C
C   C—OH
C—C

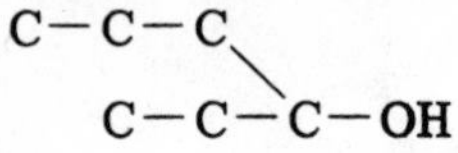

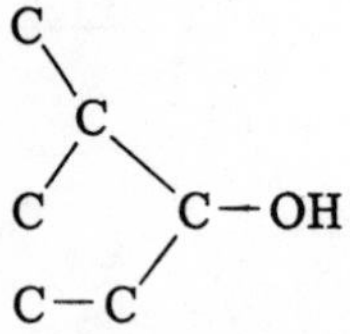

3-1-1

2:   C—C—C
          C—C—OH
          C
and   C
          C
      C
          C—C—OH
          C

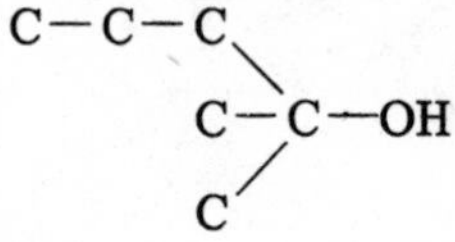

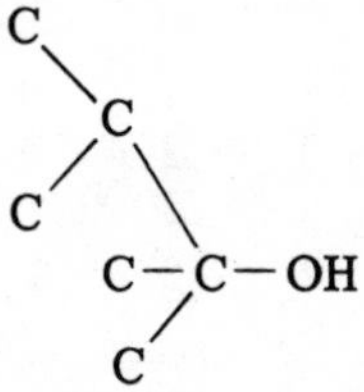

2-2-1

1:   C—C
     C—C—C—OH
          C

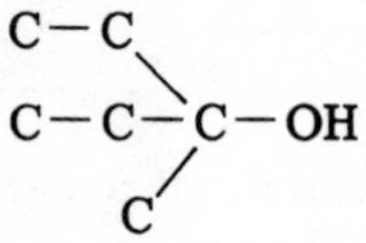

16-17   The total number of 6-carbon alcohols, therefore, is _____.

17   (i.e., 8 + 4 + 2 + 2 + 1)

16-18 The formulas of the 17 six-carbon alcohols could now be written and their systematic names worked out. Instead, we will proceed to the 7-carbon alcohols, following the same procedure as before. There is only one complication that we will solve in general terms first. In a molecule $\begin{smallmatrix} x \\ \diagdown \\ C-OH, \\ \diagup \\ y \end{smallmatrix}$ suppose x can be either group A or B, and y can be either group A or B. The number of different molecules obtainable if x and y can each be either A or B is _____ .

3

16-19 The three formulas must be _____ ,

_____ , and _____ .

$\begin{smallmatrix} A \\ \diagdown \\ C-OH \\ \diagup \\ A \end{smallmatrix}$
$\begin{smallmatrix} A \\ \diagdown \\ C-OH \\ \diagup \\ B \end{smallmatrix}$
and
$\begin{smallmatrix} B \\ \diagdown \\ C-OH \\ \diagup \\ B \end{smallmatrix}$

16-20 Now figure out the total number of 7-carbon alcohols. The number is _____ .

---

39 (If you don't get this answer the first time, work over the problem again until you solve it.)

16-21   If you give up or want to check your method, see below.

The Seven-Carbon Alcohols

Distribution of 6 carbons over three bonds of the COH group

x	y	z	No. of Alcohols
6	0	0	17 (see 16-17)
5	1	0	8 (see 16-12)
4	2	0	4 (see 16-4)
4	1	1	4 (see 16-4)
3	3	0	3 (see 16-3, 16-19)
3	2	1	2 (see 16-3)
2	2	2	1
		Total	39

This method can be used indefinitely for any class of compounds R-Y where R is a saturated alkyl group and Y is a functional group attached by a single bond.

17

Questions

(For answers see Chapter 18)

CHAPTER 1. Part A—Common Names of Saturated Aliphatic Hydrocarbons

1. Give the number of carbon atoms present in a molecule of the following:

(a) *n*-butane   (c) *n*-nonane

(b) isopentane   (d) neohexane

2. Give the number of secondary carbon atoms in

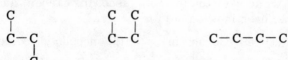

3. Write carbon skeleton structures for the three pentanes.

4. Give the number of different chemicals corresponding to the following formulas:

```
      C                C   C
      |                |   |
  C — C            C — C          C — C — C — C
      |
      C
```

5. Write the skeleton structures for

(a) ethane   (c) neopentane

(b) *n*-octane   (d) isohexane

6. Give common names corresponding to the following formulas:

(a) CH_4

(b) $C-C-C$

(c)
$$C-\underset{\underset{\textstyle C}{|}}{C}-C$$

(d) $C-C-C-C-C-C-C$

CHAPTER 1.   Part B—Common Names of Alkane Derivatives

1. Give the common name corresponding to the following formulas:

(a) CH_3Cl

(b)
$$\begin{array}{l} CH_3 \\ \quad\diagdown \\ \qquad CH-CH_2Cl \\ \quad\diagup \\ CH_3 \end{array}$$

(c) $CH_3CH_2CH_2CH_2CH_2CH_2Br$

(d)
$$\begin{array}{l} CH_3 \\ \quad\diagdown \\ CH_3-C-OH \\ \quad\diagup \\ CH_3 \end{array}$$

(e)
$$CH_3-\underset{\underset{\textstyle CH_3}{|}}{\overset{\overset{\textstyle CH_3}{|}}{C}}-CH_2-CH_2-Br$$

2. Write skeleton formulas (showing carbon, fluorine, chlorine, bromine, iodine) for

(a) ethyl bromide

(b) isopropyl fluoride

(c) *tert*-pentyl chloride

(d) neopentyl iodide

(e) isooctyl chloride

CHAPTER 2.   **Part A—The Substituted Methane System**

1.   Write skeleton formulas for

    (a)   tetraethylmethane

    (b)   diethyldimethylmethane

    (c)   triisopropylmethane

    (d)   *n*-butyldiethylisobutylmethane

2.   Name the following by the substituted methane system:

    (a) $CH_3CH_2CH_3$ (b) $(CH_3)_4C$

    (c) 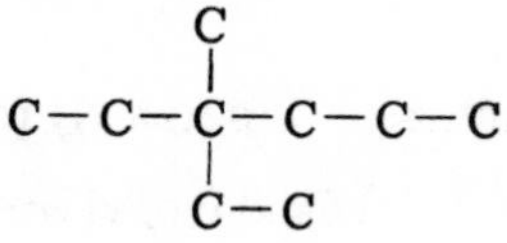
```
         C
         |
 C — C — C — C — C — C
         |
         C — C
```

    (d) 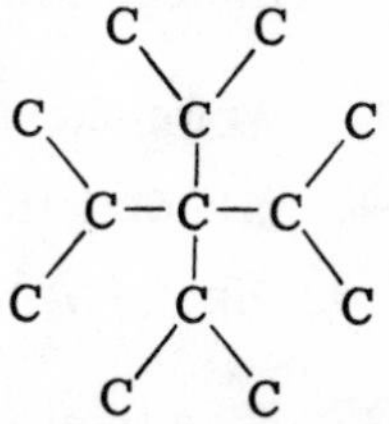
```
        C     C
         \   /
    C     C
     \    |     C
      C — C — C
     /    |     \
    C     C      C
         / \
        C   C
```

CHAPTER 2.   **Part B—The Carbinol System**

1.   Write skeleton structures (showing carbons and OH group) for the following:

    (a)   carbinol

    (b)   ethylcarbinol

    (c)   triethylcarbinol

    (d)   *n*-butyl-*tert*-butylisobutylcarbinol

2.   Name the following by the carbinol system:

(a) $(CH_3)_3COH$

(b) CH_3CH_2
     $CH_3CH_2CH_2$ $CHOH$

(c) $CH_3 - \overset{\displaystyle H}{\underset{\displaystyle OH}{C}} - CH_2CH_2CH_3$

(d) $CH_3CH_2 - \overset{\displaystyle CH_3}{\underset{\displaystyle \underset{\textstyle CH_3 \quad CH_3}{CH}}{C}} - OH$

CHAPTER 3.   Amines

1.   Write skeleton structures (C's and N) for

(a) dimethylamine

(b) tri-*n*-butylamine

(c) isopropyl-*n*-propylamine

2.   Give the names of

(a) $C - C - \overset{\displaystyle}{\underset{\displaystyle C}{N}} - C$

(b) $(CH_3)_3N$

(c) $(CH_3CH_2)_2NCH_2CH_2CH_2CH_3$

CHAPTER 4.   Systematic Names of Alkanes

1.   Write skeleton formulas for

(a) propane

(b) pentane

(c) 2-methylbutane

(d) 2,2-dimethylhexane

(e) 3-ethyl-2,3-dimethyl-5-propyloctane

2.   Write complete structural formulas (showing all atoms and bonds) for:

(a) 2-methylpentane

(b) 4-ethyl-2,4-dimethylhexane

3. Write condensed structural formulas [e.g., $CH_3(CH_2)_4CH_3$, $(CH_3)_3CH$] for

   (a) pentane

   (b) 3-ethylpentane

   (c) 2,2-dimethylpropane

4. Write the Systematic (IUPAC) names for

   (a) $CH_3CH_2CH_2CH_3$

   (b) 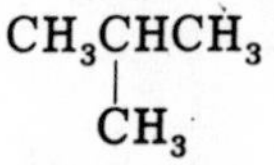 CH_3CHCH_3
            |
           CH_3

   (c) $(CH_3)_3CH$

   (d) 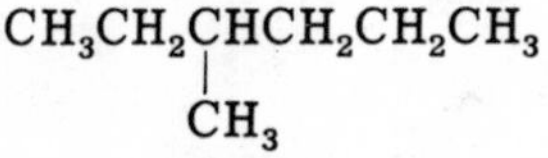 $CH_3CH_2CHCH_2CH_2CH_3$
                    |
                   CH_3

                   CH_3
                   |
   (e) 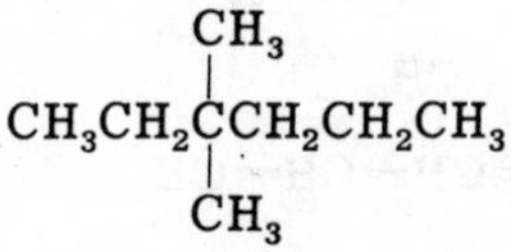 $CH_3CH_2CCH_2CH_2CH_3$
                   |
                   CH_3

                 CH_3
                |
   (f) 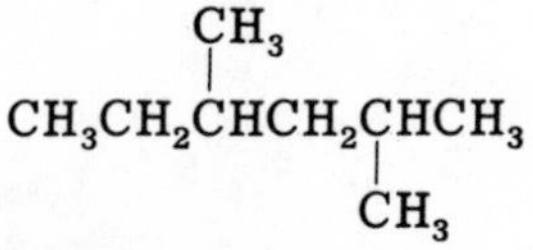 $CH_3CH_2CHCH_2CHCH_3$
                         |
                       CH_3

   (g) 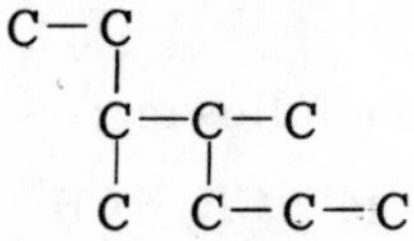
```
C—C
  |
  C—C—C
  |   |
  C   C—C—C
```

   (h) 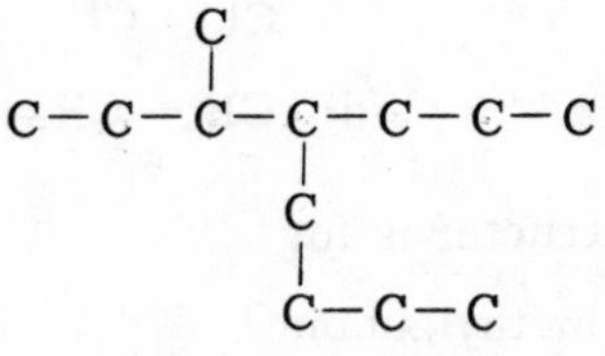
```
        C
        |
C—C—C—C—C—C—C
      |
      C
      |
      C—C—C
```

CHAPTER 5.   **Alkenes, etc.**

1.   Write the carbon skeleton formula (showing the location of the double bond) for

(a) propene

(b) 2-pentene

(c) 3-methyl-3-hexene

(d) 1,3-pentadiene

(e) cyclopentene

(f) 1,4-cyclohexadiene

(g) 2-methyl-3-heptyne

2.   Give IUPAC names for

(a) $CH_2{=}CH{-}CH_2{-}CH_3$

(b) $CH_3{-}CH{=}\underset{\underset{\displaystyle CH_3}{|}}{C}{-}CH_3$

(c) $CH_3{-}CH_2{-}\underset{\underset{\displaystyle CH{=}CH_2}{|}}{CH}{-}CH_2{-}CH_3$

(d) $CH_2{=}CH{-}CH{=}CH{-}CH{=}CH_2$

(e) $\underset{\displaystyle CH{=}CH}{\overset{\displaystyle CH{=}CH}{|\qquad|}}$

CHAPTER 6.   **Alkyl Halides**

1.   Give IUPAC names for the following:

(a) $CH_3CH_2CH_2Br$

(b) $(CH_3)_3CBr$

(c) $\underset{\underset{\displaystyle Cl}{|}}{CH_2}{-}\underset{\underset{\displaystyle Cl}{|}}{CH}{-}CH_2{-}CH_2Cl$

(d) $CH_2{-}CH{=}CH{-}CH_2Cl$

2.   Write skeleton structures for

(a) 2-chloro-2-methylbutane

(b) 2,4-dichloro-3,5-dimethyloctane

(c) 1,1,1-trichloroethane

(d) 3-bromo-1-pentene

CHAPTER 7. **Alcohols**

1. Write condensed structural formulas [e.g. $CH_3CH_2CH_2CH_2OH$, $(CH_3)_2CHOH$, etc.] for

    (a) 2-propanol

    (b) 2-methyl-2-propanol

    (c) 2,4-dimethyl-2-pentanol

    (d) 3-penten-2-ol

    (e) 2,4-hexanediol

    (f) 2-methylcyclobutanol

2. Give the IUPAC name for

    (a) $CH_3CH_2CH_2CH_2CH_2CH_2CH_2CH_2CH_2CH_2OH$

    (b) 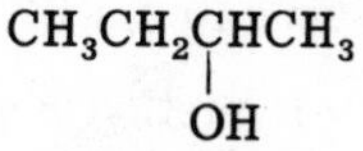 $CH_3CH_2\underset{\underset{OH}{|}}{C}HCH_3$

    (c)
$$C-C-\underset{\underset{\underset{C}{|}}{C-C-OH}}{\overset{|}{C}}-C$$

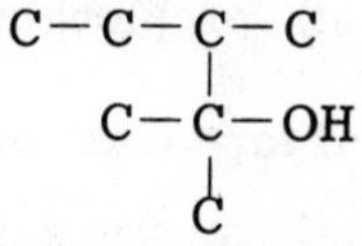

    (d) 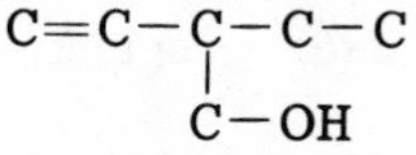
$$C=C-\underset{\underset{C-OH}{|}}{C}-C-C$$

CHAPTER 8. **Aldehydes and Ketones**

1. Name the following:

    (a) 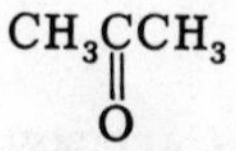 $CH_3\underset{\underset{O}{\|}}{C}CH_3$

    (b) $CH_3CH_2CH_2CH_2CH_2CH_2CHO$

    (c) 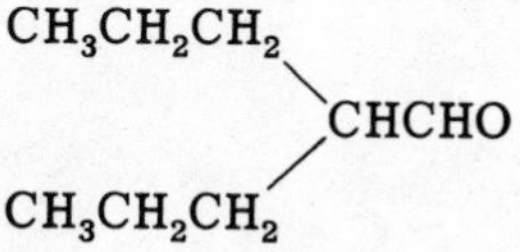 $\begin{matrix} CH_3CH_2CH_2 \\ \diagdown \\ \quad\quad CHCHO \\ \diagup \\ CH_3CH_2CH_2 \end{matrix}$

(d) CH$_3$CH$_2$CH$_2$
     CH$_3$CH$_2$CH$_2$ $\Big\rangle$C=O

(e) CH$_3$CH$_2$CCH$_2$CHCH$_2$CH$_2$CH$_3$
              ‖ |
              O   CH$_2$CH$_3$

2. Write skeleton formulas for:

   (a) pentanal   (c) 2,2,2-trichloroethanal

   (b) 4-butyl-2-octanone (d) 3-ethylhexanal

CHAPTER 9.   Common Names of Carboxylic Acids, Aldehydes, and Ketones

1. What is the number of carbon atoms in

   (a) butyric acid   (c) methyl *n*-propyl ketone

   (b) heptylic acid   (d) propionaldehyde

2. Write condensed structural formulas for

   (a) *n*-butyl ethyl ketone (c) formic acid

   (b) acetaldehyde   (d) propionic acid

3. Give the common name of each of the following:

   (a) CH$_3$CH$_2$CH$_2$CH$_2$CO$_2$H (d) CH$_3$CH$_2$COCH$_2$CH$_3$

   (b) (CH$_3$)$_2$CHCO$_2$H   (e) CH$_3$COCH(CH$_3$)$_2$

   (c) CH$_3$CH$_2$CHO   (f) CH$_3$(CH$_2$)$_4$CHO

CHAPTER 10.   Systematic Names of Carboxylic Acids

1. Give the systematic names of the following:

   (a) CH$_3$CH$_2$CO$_2$H

   (b) HCO$_2$H

(c) $CH_3CH_2CH_2CH_2CH_2CO_2H$

(d) $CH_2CH_2CO_2H$
$\quad\ \ CH_2CH_2CO_2H$

(e) $CH_3CH_2CHCH_2CHCH_2CO_2H$
$\qquad\quad CH_3 \quad\ CH_3$

(f) $CH_3-CH=CH-CO_2H$

2. Write condensed structural formulas for:

(a) butanoic acid

(b) 3-propyloctanoic acid

(c) 2-hydroxypropanoic acid

(d) 2-pentenedioic acid

(e) 2,2-diethyl-6-propylnonanoic acid

CHAPTER 11. Esters

1. Write condensed structural formulas for:

(a) ethyl acetate

(b) methyl isobutyrate

(c) isopropyl acetate

(d) *tert*-pentyl propionate

2. Give the *common* name for

(a) $CH_3-\underset{\underset{O}{\|}}{C}-OCH_2CH_2CH_2CH_3$

(b) $CH_3CH_2CO_2CH_3$

(c) $(CH_3)_2CHCO_2CH(CH_3)_2$

3. Give the IUPAC name for each of the formulas listed in Question 2.

CHAPTER 12. **Aromatic Compounds**

1.   Using the symbol 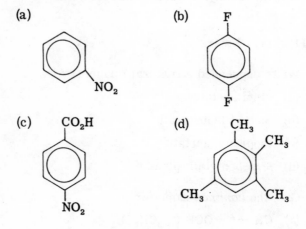 , to represent the

benzene ring (plus all hydrogens attached to the ring) write
formulas for the following:

    (a)   *m*-dibromobenzene

    (b)   1-chloro-2-nitrobenzene

    (c)   2-chloroaniline

    (d)   *p*-nitrophenol

    (e)   1-bromo-2,4-dinitrobenzene

2.   Name the following, using numbers to designate
positions where necessary.

(a)

(b)   F

(c) CO_2H

(d) CH_3

CHAPTER 13. Formal Charges

1. Give the formal charge on

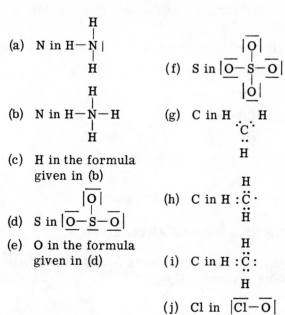

(a) N in H—N|
    with H on top and H on bottom

(b) N in H—N—H
    with H on top and H on bottom

(c) H in the formula given in (b)

(d) S in |O̅—S̲—O̅|
    with |O̅| on top

(e) O in the formula given in (d)

(f) S in |O̅—S—O̅|
    with |O̅| on top and |O̅| on bottom

(g) C in H ..C.. H with H

(h) C in H :C̈· with H

(i) C in H :C̈: with H

(j) Cl in |C̅l—O̅|

CHAPTER 14. Electronic Formulas

1. Write electronic structures for the following, using lines for both shared and unshared electron pairs, and indicating formal charges.

(a) NO_2^+ (atomic arrangement ONO)

(b) NO_2^- (atomic arrangement ONO)

(c) ClO_3^- (oxygen attached only to Cl)

(d) H_3PO_4 (arrangement $(HO)_3PO$)

CHAPTER 15.   Resonance Formulas

1.   Write all resonance structures, showing bonds, unshared electron pairs and formal charges, for each of the following:

(a)   the SO_3 molecule (all oxygens bonded to S)

(b)   chlorobenzene

(c)   *p*-nitrophenol

(d)

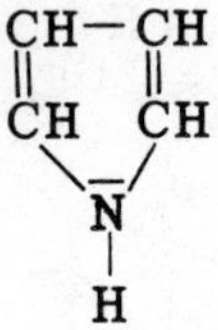

CHAPTER 16.   Isomeric Alcohols

1.   Give the skeleton structures and IUPAC names for all alcohols of formula $C_6H_{13}OH$.

18

Answers

CHAPTER 1. Part A—Common Names of Saturated Aliphatic Hydrocarbons

1. (a) 4   (b) 5   (c) 9   (d) 6
2. (a) 3   (b) 2
3. C—C—C—C—C, C—C—C—C (or C

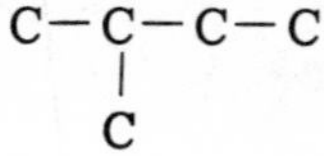

    |

    C

    ),

4. one

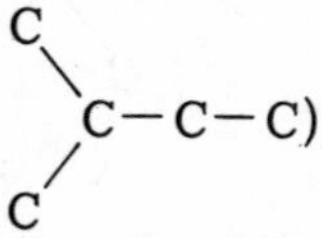

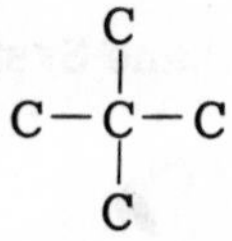

5. (a) C—C   (b) C—C—C—C—C—C—C—C

    (c) 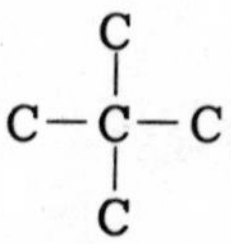   (d)

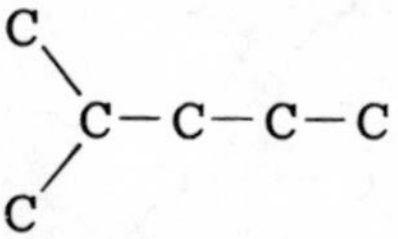

    (or C—C—C—C—C)

    |

    C

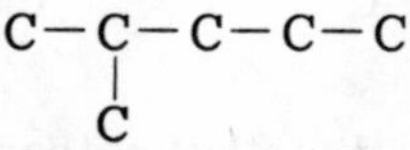

6. (a) methane   (b) propane   (c) isobutane
    (d) *n*-heptane

CHAPTER 1. Part B—Common Names of Alkane Derivatives

1. (a) methyl chloride   (d) *tert*-butyl alcohol
   (b) isobutyl chloride   (e) neohexyl bromide
   (c) *n*-hexyl bromide

2. (a) C—C—Br   (b) 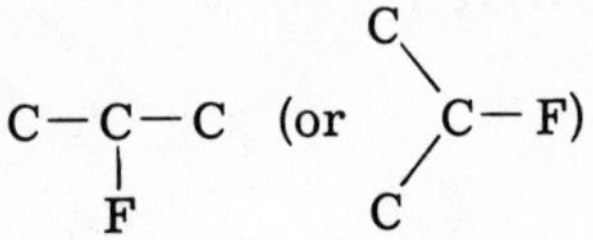 C—C—C (or $\overset{\displaystyle C}{\underset{\displaystyle C}{>}}$C—F)

                                  |
                                  F

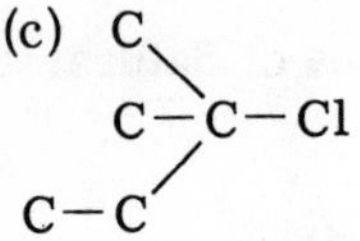

   (c) structure
   (d) 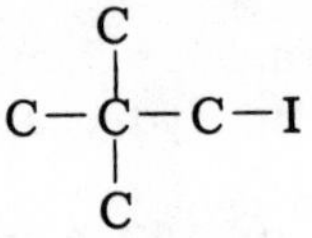 structure
   (e) 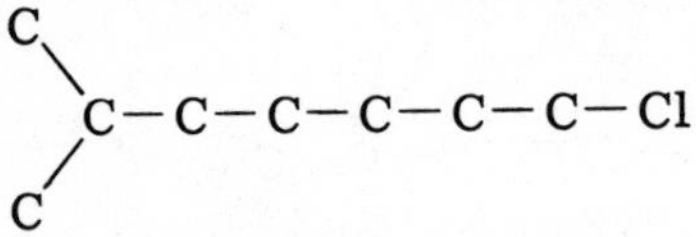 structure

CHAPTER 2. Part A—The Substituted Methane System

1. (a) 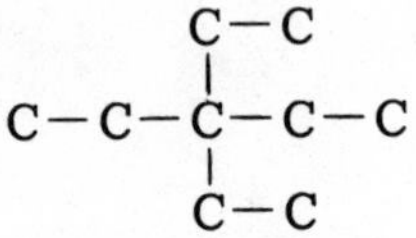 structure   (b) 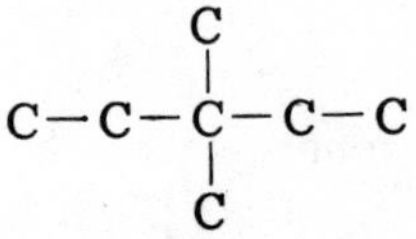 structure
   (c) 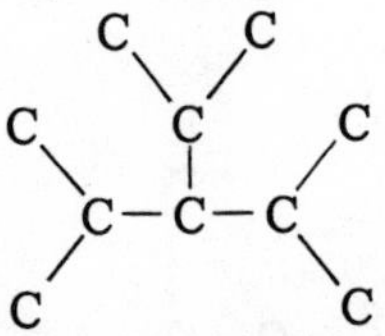 structure   (d) 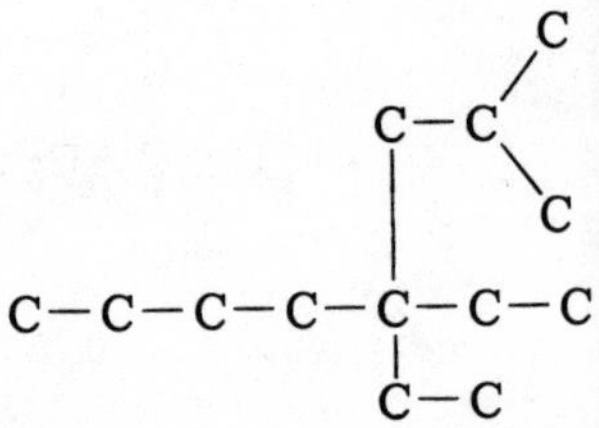 structure

2. (a) dimethylmethane
   (b) tetramethylmethane
   (c) diethylmethyl-*n*-propylmethane
   (d) tetraisopropylmethane

CHAPTER 2. **Part B—The Carbinol System**

1. (a) C—OH
   (b) C—C—C—OH
   (c)
```
   C—C
        \
   C—C—C—OH
        /
   C—C
```
   (d)
```
              C
             /
      C—C
       |       \
       |        C
       |
                     C
                    /
C—C—C—C—C—C—C
          |        \
          OH        C
```

2. (a) trimethylcarbinol
   (b) ethyl-*n*-propylcarbinol
   (c) methyl-*n*-propylcarbinol
   (d) ethylisopropylmethylcarbinol

CHAPTER 3. **Amines**

1. (a) C—N—C
   (b)
```
                C—C—C—C
                |
   C—C—C—C—N—C—C—C—C
```
   (c)
```
                      C
                     /
   C—C—C—N—C
                     \
                      C
```

2. (a) ethyldimethylamine
   (b) trimethylamine
   (c) *n*-butyldiethylamine

CHAPTER 4. **Systematic Names of Alkanes**

1. (a) C—C—C
   (b) C—C—C—C—C
   (c)
```
   C—C—C—C
       |
       C
```

(d)

$$
\begin{array}{c}
C \\
| \\
C-C-C-C-C-C \\
| \\
C
\end{array}
$$

(e)

$$
\begin{array}{c}
C-C C-C-C \\
| | \\
C-C-C-C-C-C-C-C \\
| | \\
C C
\end{array}
$$

2. (a)

$$
\begin{array}{c}
H \\
| \\
H-C-H \\
| \\
H H H H \\
| | | | \\
H-C-C-C-C-C-H \\
| | | | | \\
H H H H H
\end{array}
$$

(b)

$$
\begin{array}{c}
H \\
| \\
H H-C-H \\
| | \\
H-C-H \\
| \\
H H H H \\
| | | | \\
H-C-C-C-C-C-C-H \\
| | | | | \\
H H H H H \\
H \\
| \\
H-C-C-H \\
| | \\
H H
\end{array}
$$

3. (a) $CH_3(CH_2)_3CH_3$
    (b) $(CH_3CH_2)_3CH$ or $(CH_3CH_2)_2CHCH_2CH_3$ or
    $$
    \begin{array}{c}
    CH_3CH_2CHCH_2CH_3 \\
    | \\
    CH_2CH_3
    \end{array}
    $$
    (c) $(CH_3)_4C$ or
    $$
    \begin{array}{c}
    CH_3 \\
    | \\
    CH_3CCH_3 \\
    | \\
    CH_3
    \end{array}
    $$

4. (a) butane   (e) 3,3-dimethylhexane
    (b) 2-methylpropane (f) 2,4-dimethylhexane
    (c) 2-methylpropane (g) 3,4-dimethylheptane
    (d) 3-methylhexane (h) 3-methyl-4-propyloctane

CHAPTER 5. **Alkenes, etc.**

1. (a) C=C—C (or C—C=C)

    (b) C—C=C—C—C

    (c) C—C—C=C—C—C
          |
          C

    (d) C=C—C=C—C

    (e)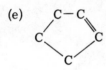

    (f) 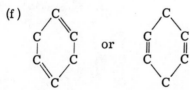 or

    (g) C—C—C≡C—C—C—C
          |
          C

2. (a) 1-butene   (d) 1,3,5-hexatriene
    (b) 2-methyl-2-butene (e) 1,3-cyclobutadiene
    (c) 3-ethyl-1-pentene

CHAPTER 6. **Alkyl Halides**

1. (a) 1-bromopropane
    (b) 2-bromo-2-methylpropane
    (c) 1,2,4-trichlorobutane
    (d) 1-chloro-2-butene

2. (a)

$$\begin{array}{c} C \\ | \\ C-C-C-C \\ | \\ Cl \end{array}$$

(b)

$$\begin{array}{c} Cl \quad C \quad Cl \quad C \\ | \quad | \quad | \quad | \\ C-C-C-C-C-C-C-C \end{array}$$

(c)

$$\begin{array}{c} Cl \\ | \\ Cl-C-C \\ | \\ Cl \end{array}$$

(d)

$$\begin{array}{c} C=C-C-C-C \\ | \\ Br \end{array}$$

CHAPTER 7.   Alcohols

1. (a) $(CH_3)_2CHOH$ or CH_3CHCH_3
         $\qquad\qquad\qquad\qquad\quad |$
         $\qquad\qquad\qquad\qquad\quad OH$

(b) $(CH_3)_3COH$ or

$$\begin{array}{c} CH_3 \\ | \\ CH_3CCH_3 \\ | \\ OH \end{array}$$

(c)

$$\begin{array}{c} OH \\ | \\ CH_3CCH_2CHCH_3 \\ | \qquad | \\ CH_3 \quad CH_3 \end{array}$$ or $(CH_3)_2C(OH)CH_2CH(CH_3)_2$

(d) $CH_3CH-CH=CH-CH_3$ or $CH_3CH(OH)CH=CHCH_3$
         $\qquad |$
         $\qquad OH$

(e) $CH_3CHCH_2CHCH_2CH_3$
         $\qquad |\qquad\quad |$
         $\qquad OH\quad\; OH$

(f)

$$\begin{array}{c} CH_3 \\ | \\ CH-CHOH \\ | \\ CH_2-CH_2 \end{array}$$

2. (a) 1-decanol   (c) 2,3-dimethyl-2-pentanol
    (b) 2-butanol   (d) 2-ethyl-3-buten-1-ol

CHAPTER 8.   Aldehydes and Ketones

1. (a) propanone (or 2-propanone)
    (b) heptanal
    (c) 2-propylpentanal
    (d) 4-heptanone
    (e) 5-ethyl-3-octanone

2. (a) C—C—C—C—CHO

    (b) C—C—C—C—C—C—C—C
              ‖ |
              O   C—C—C—C

    (c)   Cl
               |
          Cl—C—CHO
               |
               Cl

    (d) C—C—C—C—C—CHO
               |
               C—C

CHAPTER 9.   Common Names of Carboxylic Acids, Aldehydes, and Ketones

1. (a) 4   (b) 7   (c) 5   (d) 3

2. (a) $CH_3CH_2CH_2CH_2COCH_2CH_3$ or
         $CH_3(CH_2)_3COCH_2CH_3$ or $CH_3(CH_2)_3\overset{\text{O}}{\underset{\|}{C}}CH_2CH_3$

    (b) CH_3CHO
    (c) HCO_2H
    (d) $CH_3CH_2CO_2H$

3. (a) valeric acid   (d) diethyl ketone
    (b) isobutyric acid   (e) isopropyl methyl ketone
    (c) propionaldehyde   (f) caproaldehyde

CHAPTER 10. Systematic Names of Carboxylic Acids

1. (a) propanoic acid   (d) hexanedioic acid
    (b) methanoic acid   (e) 3,5-dimethylheptanoic acid
    (c) hexanoic acid   (f) 2-butenoic acid

2. (a) $CH_3(CH_2)_2CO_2H$ or $CH_3CH_2CH_2CO_2H$

    (b) $CH_3CH_2CH_2CH_2CH_2CHCH_2CO_2H$
    $$\qquad\qquad\qquad\qquad\quad |$$
    $$\qquad\qquad\qquad\qquad\ CH_2$$
    $$\qquad\qquad\qquad\qquad\ CH_2CH_3$$

    (c) CH_3CHCO_2H
    $$\quad\ |$$
    $$\quad\ OH$$

    (d) $CH{=}CHCH_2CO_2H$
    $$|$$
    $$CO_2H$$

    (e) CH_2CH_3
    $$CH_3CH_2CH_2CHCH_2CH_2CH_2CCO_2H$$
    $$\qquad\qquad\ |\qquad\qquad\qquad\ |$$
    $$\qquad\ CH_2CH_2CH_3\quad CH_2CH_3$$

CHAPTER 11. Esters

1. (a) $CH_3CO_2CH_2CH_3$

    (b) $(CH_3)_2CHCO_2CH_3$

    (c) $CH_3CO_2CH(CH_3)_2$

    (d) CH_3
    $$\qquad\qquad\qquad\quad |$$
    $$CH_3CH_2CO_2C{-}CH_3$$
    $$\qquad\qquad\qquad\quad |$$
    $$\qquad\qquad\qquad CH_2CH_3$$

2. (a) *n*-butyl acetate
    (b) methyl propionate
    (c) isopropyl isobutyrate

3. (a) butyl ethanoate
    (b) methyl propanoate

CHAPTER 12. Aromatic Compounds

1. (a) or 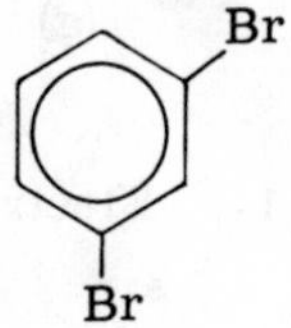 etc.

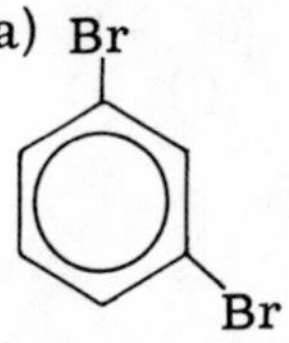

(b) or 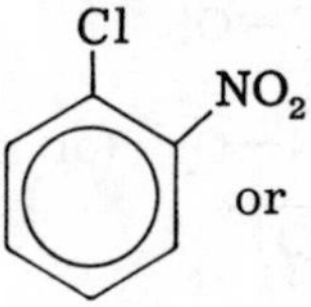 or 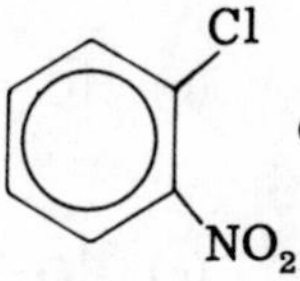 etc.

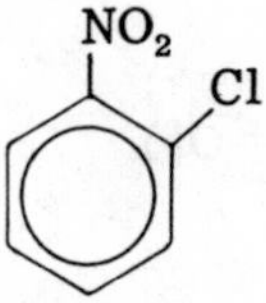

(c) (with variations as in (b))

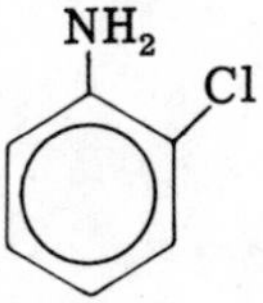

(d) etc.

(e) or 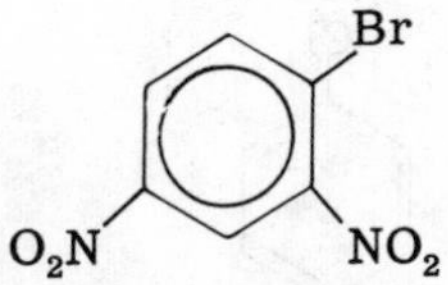 etc.

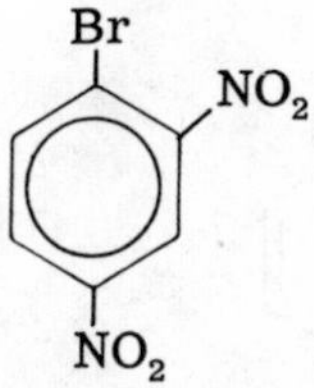

2. (a) nitrobenzene
   (b) 1,4-difluorobenzene
   (c) 4-nitrobenzoic acid
   (d) 1,2,3,5-tetramethylbenzene

CHAPTER 13.   Formal Charges

1. (a) 0   (b) +1   (c) 0   (d) +1   (e) −1
    (f) +2   (g) +1   (h) 0   (i) −1   (j) 0

CHAPTER 14.   Electronic Formulas

1. (a) $|\overline{O}=\overset{\oplus}{N}=\overline{O}|$

   (b) $|\overline{O}=\overline{N}-\overset{\ominus}{\underline{O}}|$ (or $|\overset{\ominus}{\underline{O}}-\overline{N}=\overline{O}|$)

   (c) $\overset{\ominus}{|}\overline{O}-\underset{\oplus 2}{\underline{C}l}-\overline{O}|^{\ominus}$ with $|\overline{O}|^{\ominus}$ above

   (d) H$-\overline{O}-\overset{\oplus}{P}-\overline{O}|^{\ominus}$ with $|\overline{O}-H$ above and $|\underline{O}-H$ below

CHAPTER 15.   Resonance Formulas

1. (a), (b) [resonance structures]

(c)

Plus mirror image of ring, plus mirror image of nitro group, plus mirror image of both.

Plus mirror image of ring, plus mirror image of nitro group, plus mirror image of both.

Plus mirror image of nitro group.

(d)

CHAPTER 16.   Isomeric Alcohols

2-2-1   (1) 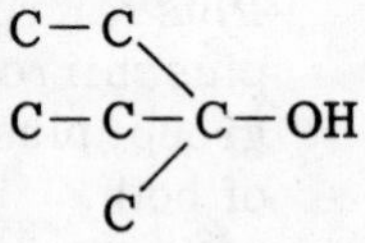   C—C
                C—C—C—OH
                   C

3-methyl-3-pentanol

3-2-0   (2) 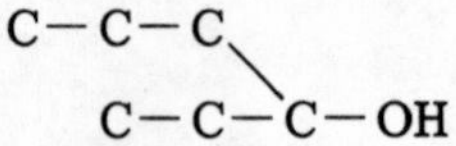   C—C—C
                  C—C—C—OH

3-hexanol

(3) 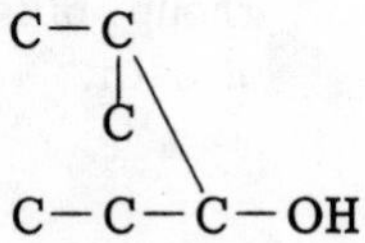   C—C
            C

C—C—C—OH

2-methyl-3-pentanol

3-1-1   (4) 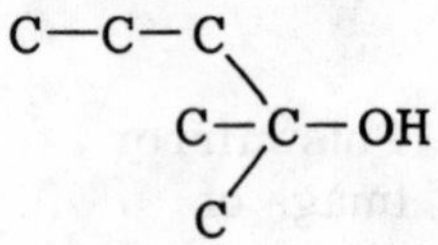   C—C—C
                  C—C—OH
                   C

2-methyl-2-pentanol

(5) 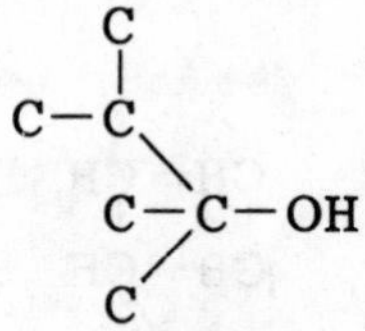   C
          C—C
             C—C—OH
           C

2,3-dimethyl-2-butanol

4-1-0   (6) 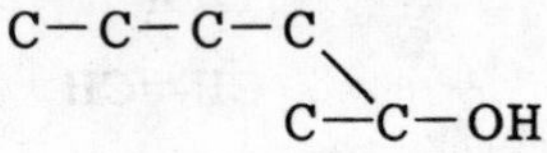   C—C—C—C
                     C—C—OH

2-hexanol

(7)

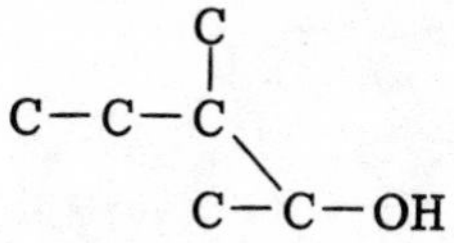

```
        C
        |
C — C — C
         \
        C — C — OH
```

3-methyl-2-pentanol

(8)

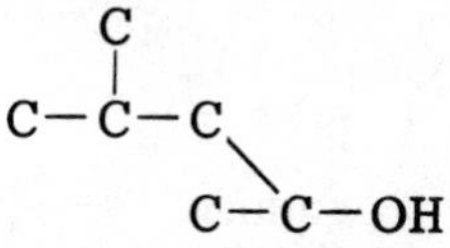

```
    C
    |
C — C — C
         \
        C — C — OH
```

4-methyl-2-pentanol

(9)

```
    C
    |
C — C
    |
    C
    |
    C — C — OH
```

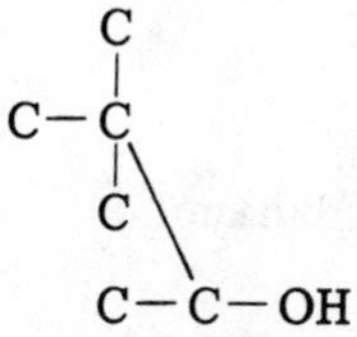

3,3-dimethyl-2-butanol

5-0-0 (10)

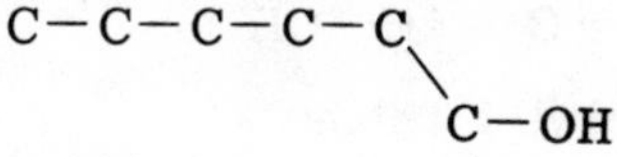

```
C — C — C — C — C
                 \
                 C — OH
```

1-hexanol

(11)

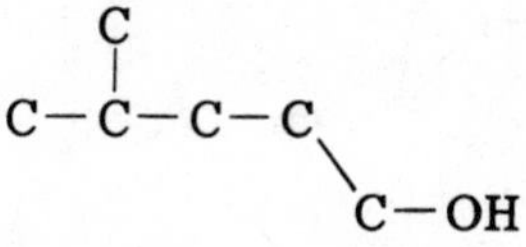

```
    C
    |
C — C — C — C
             \
             C — OH
```

4-methyl-1-pentanol

(12)

```
        C
        |
C — C — C — C
             \
             C — OH
```

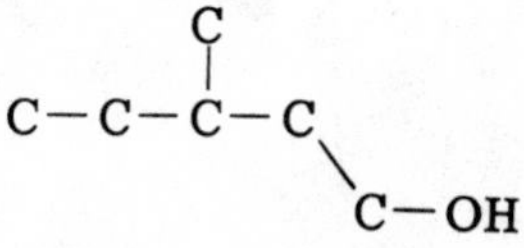

3-methyl-1-pentanol

(13)

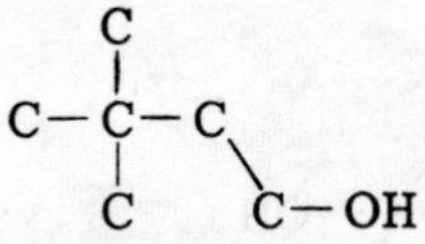

```
        C
        |
    C — C — C
        |
        C       C — OH
```

3,3-dimethyl-1-butanol

(14)

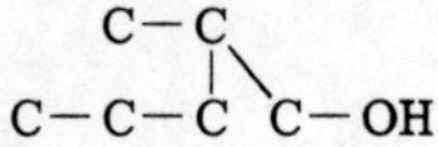

```
        C — C
            |
    C — C — C   C — OH
```

2-methyl-1-pentanol

(15)

```
    C — C
    |
    C — C       C — OH
    |
    C
```

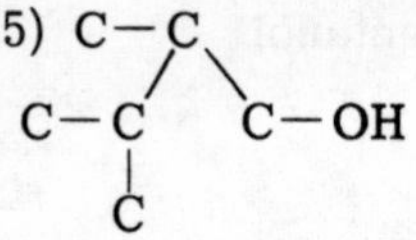

2,3-dimethyl-1-butanol

(16)

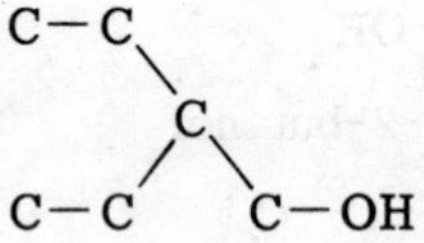

```
    C — C
        \
         C
        /    \
    C — C     C — OH
```

2-ethyl-1-butanol

(17)

```
    C — C
        \
         C — C
        /     \
    C          C — OH
```

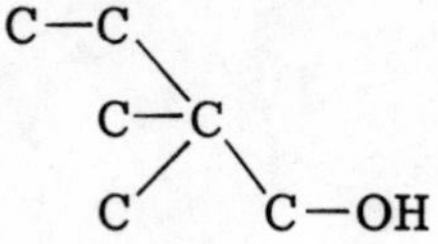

2,2-dimethyl-1-butanol

Bibliography

1. The "Definitive Rules for the Nomenclature of Organic Chemistry" were adopted unanimously by the Commission on Nomenclature and by the Council of the International Union of Pure and Applied Chemistry (IUPAC) in 1957 in Paris. They were published for the Union as "I.U.P.A.C. Nomenclature of Organic Chemistry, 1957" by Butterworths Scientific Publications in 1958. As new rules are tentatively adopted they are published in the Bulletin of the Union. When made 'definitive' they appear in the Union's journal *Pure and Applied Chemistry*. The IUPAC rules appeared, with comments, in *J. Am. Chem. Soc.* **82**, 5545-5584 (1960). Reprints are available from Chemical Abstracts Service.

2. Extracts from the IUPAC rules have been reprinted in "Handbook of Chemistry and Physics," Robert C. Weast, Editor, The Chemical Rubber Co., Cleveland, Ohio, 45th ed., 1964, pp. C-1 ff. Variations from these rules, employed by Chemical Abstracts, are pointed out.

3. "The Naming and Indexing of Chemical Compounds from Chemical Abstracts," Chemical Abstracts Service, Columbus, Ohio, 1962, 98 pp.

This is a reprint of the "Introduction to the Subject Index to Volume 56 (January to June, 1962)" of Chemical Abstracts. It presents the rules followed in Chemical Abstracts and points out variations from IUPAC practice.

4. Cahn, R. S., "An Introduction to Chemical Nomenclature," Butterworth Scientific Publications, London, 1959, 96 pp. The author is Editor for the Chemical Society, London, and points out British usage.

5. Hurd, C. D., *J. Chem. Educ.* **38**, 43-7 (1961). General Discussion of Organic Nomenclature.

6. Crane, E. J., *J. Chem. Educ.* **8**, 1335 (1931). An earlier discussion of proper nomenclature usage.

7. See also bibliography on page 67N of Reference 3 above.

Appendix

A TABULATION OF USEFUL NAMES

1. Alkanes: IUPAC Names

Methane	CH_4	Pentadecane	$C_{15}H_{32}$
Ethane	C_2H_6	Hexadecane	$C_{16}H_{34}$
Propane	C_3H_8	Heptadecane	$C_{17}H_{36}$
Butane	C_4H_{10}	Octadecane	$C_{18}H_{38}$
Pentane	C_5H_{12}	Nonadecane	$C_{19}H_{40}$
Hexane	C_6H_{14}	Eicosane	$C_{20}H_{42}$
Heptane	C_7H_{16}	Heneicosane	$C_{21}H_{44}$
Octane	C_8H_{18}	Docosane	$C_{22}H_{46}$
Nonane	C_9H_{20}	Tricosane	$C_{23}H_{48}$
Decane	$C_{10}H_{22}$	Tetracosane	$C_{24}H_{50}$
Undecane	$C_{11}H_{24}$	Pentacosane	$C_{25}H_{52}$
Dodecane	$C_{12}H_{26}$	Triacontane	$C_{30}H_{62}$
Tridecane	$C_{13}H_{28}$	Tetracontane	$C_{40}H_{82}$
Tetradecane	$C_{14}H_{30}$	Pentacontane	$C_{50}H_{102}$

2. Hydrocarbons: Some Common and IUPAC Names

Alkanes	Common	IUPAC
CH_3CH_3	Ethane	Ethane
$CH_3CH_2CH_2CH_3$	*n*-Butane	Butane

CH_3CHCH_3
    $\vert$
   CH_3
	Isobutane	Isobutane*

CH_3
    $\vert$
CH_3-C-CH_3
    $\vert$
    CH_3
	Neopentane	Neopentane*

$(CH_3)_2CHCH_2CH_2CH_3$	Isoheptane	2-Methylhexane

Alkenes		
$CH_2{=}CH_2$	Ethylene	Ethylene
$CH_2{=}C{=}CH_2$	Allene	Allene
$CH_3-CH{=}CH_2$	Propylene	Propene
$CH_3CH{=}CHCH_3$	*sym*-Dimethyl- ethylene	2-Butene

$CH_2{=}C-CH{=}CH_2$
     $\vert$
    CH_3
	Isoprene	Isoprene†

Alkynes		
$CH{\equiv}CH$	Acetylene	Acetylene
$CH_3-C{\equiv}CH$		Propyne

---

*For unsubstituted hydrocarbons only. Chemical Abstracts uses 2-Methylpropane and 2,2-Dimethylpropane, respectively.

† For the unsubstituted compound only.

Cyclic hydrocarbons	Common	IUPAC
$\begin{array}{c} CH_2-CH_2 \\ \| \qquad \| \\ CH_2-CH_2 \end{array}$	Cyclobutane	Cyclobutane
$\begin{array}{c} CH_2-CH_2 \\ CH \qquad\qquad CH \\ CH-CH \end{array}$		1,3-cyclo-hexadiene

3. Monovalent hydrocarbon groups: Common and IUPAC Names

	Common	IUPAC
CH_3-	Methyl	Methyl
C_2H_5-	Ethyl	Ethyl
$CH_3CH_2CH_2-$	*n*-Propyl	Propyl
$CH_3\overset{\|}{C}HCH_3$	Isopropyl	Isopropyl*
$CH_3CH_2CH_2CH_2-$	*n*-Butyl	Butyl
$CH_3CH_2\overset{\|}{C}HCH_3$	*sec*-Butyl	*sec*-Butyl*
$CH_3-\overset{\|}{\underset{CH_3}{C}}-CH_3$	*tert*-Butyl	*tert*-Butyl*
$CH_3-\overset{\|}{\underset{CH_3}{C}}H-CH_2-$	Isobutyl	Isobutyl*
$CH_3CH_2CH_2CH_2CH_2-$	*n*-Amyl or *n*-Pentyl	Pentyl
$CH_3-\overset{\|}{\underset{CH_3}{C}}H-CH_2-CH_2-$	Isoamyl or isopentyl	Isopentyl*

*For the unsubstituted radical only.

	Common	IUPAC
$CH_3-\overset{\overset{\displaystyle CH_3}{\vert}}{\underset{\underset{\displaystyle CH_3}{\vert}}{C}}-CH_2-$	Neopentyl	Neopentyl*
$(CH_3)_2CH(CH_2)_5-$	Isooctyl	6-Methylheptyl
$CH_2=CH-$	Vinyl	Vinyl
$CH_2=CH-CH_2-$	Allyl	Allyl
C_6H_5-	Phenyl	Phenyl
$C_6H_5CH_2-$	Benzyl	Benzyl

4. Carboxylic Acids and Aldehydes: Common Names

HCO_2H	Formic acid
CH_3CO_2H	Acetic acid
$CH_3CH_2CO_2H$	Propionic acid
$CH_3(CH_2)_2CO_2H$	Butyric acid
$CH_3(CH_2)_3CO_2H$	Valeric acid
$CH_3(CH_2)_4CO_2H$	Caproic acid
$CH_3(CH_2)_5CO_2H$	n-Heptylic acid
$CH_3(CH_2)_6CO_2H$	Caprylic acid
$CH_2(CH_2)_7CO_2H$	n-Nonylic acid
$CH_3(CH_2)_8CO_2H$	Capric acid
$CH_3(CH_2)_{10}CO_2H$	Lauric acid
$CH_3(CH_2)_{14}CO_2H$	Palmitic acid
$CH_3(CH_2)_{16}CO_2H$	Stearic acid

---

*For the unsubstituted radical only.

$CH_2=CHCO_2H$	Acrylic acid
$CH_3(CH_2)_7CH=CH(CH_2)_7CO_2H$	Oleic acid
HCHO	Formaldehyde
CH_3CHO	Acetaldehyde
CH_3CH_2CHO	Propionaldehyde
$CH_3(CH_2)_2CHO$	Butyraldehyde
$CH_3(CH_2)_3CHO$	Valeraldehyde

5. Dicarboxylic Acids

$$\begin{array}{l} CO_2H \\ | \\ (CH_2)_n \\ | \\ CO_2H \end{array}$$

n =	
0	Oxalic acid
1	Malonic acid
2	Succinic acid
3	Glutaric acid
4	Adipic acid
5	Pimelic acid
6	Suberic acid
7	Azelaic acid
8	Sebacic acid

$$\begin{array}{l} H-C-CO_2H \\ \;\;\;\; \| \\ H-C-CO_2H \end{array}$$ Maleic acid

$$\begin{array}{l} H-C-CO_2H \\ \;\;\;\; \| \\ HO_2C-C-H \end{array}$$ Fumaric acid

6. Acids and Aldehydes: Derived System

Cl_2CHCO_2H	Dichloroacetic acid
$(C_2H_5)_2CHCO_2H$	Diethylacetic acid

$[(CH_3)_3C]_3CCO_2H$	Tri-*tert*-butylacetic acid
CH_3CHO	Acetaldehyde
Cl_3CCHO	Trichloroacetaldehyde
$(CH_3)_3CCHO$	Trimethylacetaldehyde

7. Substituted Acids: Greek Letter and IUPAC System

	Greek Letter	IUPAC
$ClCH_2CO_2H$	Chloroacetic acid	Chloroacetic acid or Chloroethanoic acid
$CH_3\overset{\underset{\mid}{Cl}}{C}HCO_2H$	α-Chloropropionic acid	2-Chloropropionic acid or 2-chloropropanoic acid
$\overset{\beta}{C}H_2\overset{\alpha}{C}H_2CO_2H$ $\underset{\mid}{OH}$	β-Hydroxypropionic acid	3-Hydroxypropionic acid or 3-hydroxypropanoic acid
$\overset{\gamma}{C}H_2\overset{\beta}{C}H_2\overset{\alpha}{C}H_2CO_2H$ $\underset{\mid}{NH_2}$	γ-Aminobutyric acid	4-Aminobutyric acid or 4-aminobutanoic acid
$CH_2(CH_2)_nCO_2H$ $\underset{\mid}{Cl}$	An ω-chloro acid	

8. Derivatives of Carboxylic Acids

	Common Name	IUPAC Name
Acid		
$CH_3-\underset{\underset{O}{\parallel}}{C}-OH$	Acetic acid	Acetic acid or Ethanoic acid

	Common Name	IUPAC Name

Acid Chloride

$CH_3-\overset{\underset{\|}{O}}{C}-Cl$ Acetyl chloride   Acetyl chloride or Ethanoyl chloride

Acid Anhydride

$CH_3-C=O$
$\qquad\qquad O$
$CH_3-C=O$ Acetic anhydride   Acetic anhydride or Ethanoic anhydride

Ester

$CH_3-\overset{\underset{\|}{O}}{C}-OCH_2CH_2CH_3$ *n*-Propyl acetate   Propyl acetate or Propyl ethanoate

Amides

$CH_3-\overset{\underset{\|}{O}}{C}-NH_2$ Acetamide   Acetamide or Ethanamide

$CH_3-\overset{\underset{\|}{O}}{C}-NHCH_3$ *N*-Methylacetamide   *N*-Methylacetamide or *N*-Methyl-ethanamide

$CH_3-\overset{\underset{\|}{O}}{C}-N(CH_3)_2$ *N*,*N*-Dimethylacetamide   *N*,*N*-Dimethylacetamide or *N*,*N*-Dimethylethanamide

Nitrile

$CH_3CH_2CH_2CN$ Butyronitrile   Butyronitrile or Butanenitrile

9. Alcohols

	Common	IUPAC
$CH_3CH_2CH_2OH$	*n*-Propyl alcohol	1-Propanol
$CH_3-\underset{\underset{OH}{\|}}{CH}-\underset{\underset{OH}{\|}}{CH_2}$	Propylene glycol	1,2-Propanediol
$(CH_3CH_2)_3COH$	Triethylcarbinol	3-Ethyl-3-pentanol

10. Amines

	Common	IUPAC
$CH_3CH_2CH_2CH_2NH_2$	*n*-Butylamine	Butylamine
$CH_3NHCH_2CH_3$	Ethylmethyl-amine	*N*-Methylethyl-amine
$CH_3CH_2\overset{\displaystyle \mid}{C}HCH_3$ 　　　$NHCH_2CH_3$	*sec*-Butylethyl-amine	*N*-Ethyl-1-methyl-propylamine

11. Ethers and Epoxides

	Common	IUPAC
CH_3OCH_3	Dimethyl ether	Dimethyl ether or Methoxymethane
$CH_3OC_2H_5$	Ethyl methyl ether	Ethyl methyl ether or Methoxyethane
$(CH_3CH_2CH_2CH_2)_2O$	Di-*n*-butyl ether	Dibutyl ether or 1-butoxybutane
$\underset{\displaystyle O}{CH_2-CH_2}$	Ethylene oxide	Epoxyethane
$\underset{\displaystyle O}{CH_3CH-CHCH_2CH_3}$		2,3-Epoxypentane

12. Some Common Names of Aromatic Compounds

$C_6H_5CH_3$	Toluene
$p\text{-}C_6H_4(CH_3)_2$	*p*-Xylene
C_6H_5OH	Phenol
$C_6H_5NH_2$	Aniline
C_6H_5CHO	Benzaldehyde

$C_6H_5CO_2H$	Benzoic acid
$C_6H_5CH{=}CH_2$	Styrene
$C_6H_5CH{=}CHCO_2H$	Cinnamic acid
$o\text{-}C_6H_4(OH)_2$	Catechol
$m\text{-}C_6H_4(OH)_2$	Resorcinol
$p\text{-}C_6H_4(OH)_2$	Hydroquinone
$m\text{-}CH_3C_6H_4OH$	m-Cresol
$p\text{-}BrC_6H_4OH$	p-Bromophenol
$o\text{-}HOC_6H_4CO_2H$	Salicylic acid
$o\text{-}C_6H_4(CO_2H)_2$	Phthalic acid
$m\text{-}C_6H_4(CO_2H)_2$	Isophthalic acid
$p\text{-}C_6H_4(CO_2H)_2$	Terephthalic acid

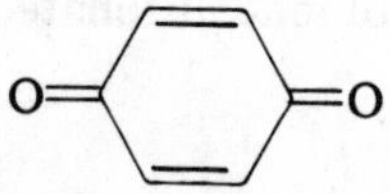

p-Benzoquinone 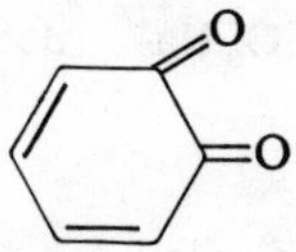 o-Benzoquinone

$C_6H_5-N{=}N-C_6H_5$ $C_6H_5-\underset{H}{N}-\underset{H}{N}-C_6H_5$

Azobenzene   Hydrazobenzene

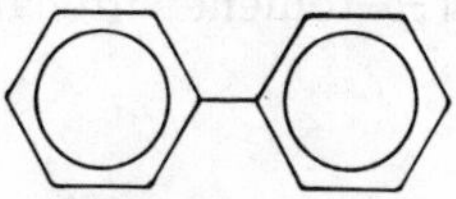

Biphenyl 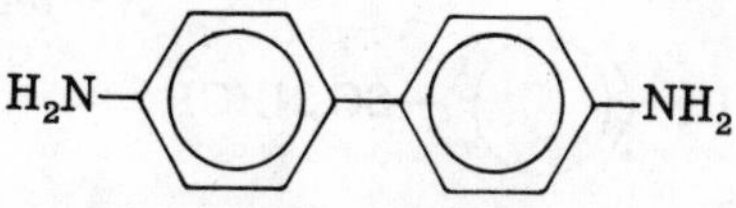   Benzidine

13. Common Names of Aromatic Sulfonic Acids and Derivatives

$C_6H_5SO_3H$ Benzenesulfonic acid

SO$_3$H / CH$_3$ *m*-Toluenesulfonic acid

SO$_3$H / NO$_2$ *o*-Nitrobenzenesulfonic acid

CH$_3$—SO$_3^-$Na$^+$ Sodium *p*-toluenesulfonate

CH$_3$—SO$_2$Cl   *p*-Toluenesulfonyl chloride

—SO$_2$NH$_2$ Benzenesulfonamide

N-Methyl-*p*-toluenesulfonamide

CH$_3$—SO$_2$NHCH$_3$

14. Benzene and Some Fused Ring Systems

Benzene

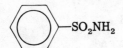

Naphthalene

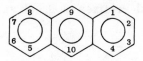

Anthracene

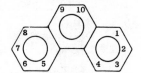

Phenanthrene

15. Some Heterocyclic Compounds

Quinoline

Isoquinoline

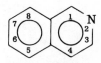

Pyridine

Pyrrole   Furan   Thiophene

16. Some Numerical Prefixes of Greek and Latin Origin

	Greek	Latin
$\frac{1}{2}$	hemi	semi
1	mono	uni
$1\frac{1}{2}$		sesqui
2	di	bi
3	tri	tri, ter
4	tetra	quadri, quater
5	penta	quinque
6	hexa	sexi
7	hepta	septi
8	octa	
9	ennea	nona
10	deca	
11	hendeca	undeca
12	dodeca	

17. Greek Alphabet

α	alpha	ν	nu
β	beta	ξ	xi
γ	gamma	o	omicron
δ	delta	π	pi
ϵ	epsilon	ρ	rho
ζ	zeta	σ	sigma
η	eta	τ	tau
θ	theta	υ	upsilon
ι	iota	ϕ	phi
κ	kappa	χ	chi
λ	lambda	ψ	psi
μ	mu	ω	omega

Index